XXXVI IULTCS Congress 2021

Greening the Leather Value Chain

Addis Ababa, Ethiopa
3-5 November 2021

ISBN: 978-1-7138-4257-6

Printed from e-media with permission by:

Curran Associates, Inc.
57 Morehouse Lane
Red Hook, NY 12571

Some format issues inherent in the e-media version may also appear in this print version.

Printed with permission by Curran Associates, Inc. (2022)

For permission requests, please contact International Union of Leather Technologists and Chemists Societies (IULTCS)
at the address below.

International Union of Leather Technologists and Chemists Societies (IULTCS)
Postfach 505
CH-4016 Basel
Switzerland

www.iultcs.org

Additional copies of this publication are available from:

Curran Associates, Inc.
57 Morehouse Lane
Red Hook, NY 12571 USA
Phone: 845-758-0400
Fax: 845-758-2633
Email: curran@proceedings.com
Web: www.proceedings.com

TABLE OF CONTENTS

XXXVI IULTCS Congress "Greening the Leather Value Chain".

Patagonian fish skin tanning processes

Gabriel Fabián Trachter
Secretaria de Pesca de la provincia de Chubut, Argentina +54 9 11 4160-9151
fabiantrachter@hotmail.com

Elisa Palomino
Central Saint Martins, University of the Arts, London, UK +44 7404430210
e.palomino@csm.arts.ac.uk

Gustavo Adrian Defeo
Ars Tinctoria s.r.l. analytical laboratory, Santa Croce sull'Arno, Italy +39 348 8924726
g.defeo@arstinctoria.it

Abstract
The ancient tradition of using fish skin to create clothing and accessories is shared by several coastal Arctic societies as part of their subsistence lifestyle depending on aquatic resources for nourishment and clothing.

Antarctica has no population, but the Tehuelches, Selk'nam, Yámanas and Alakaluf are some of the world's southernmost Indigenous Peoples closest to Antarctica. They live in the Patagonian region of southern Argentina, Chile, including Tierra del Fuego islands. They fish and hunt animals whose migrations in the Antarctic are a critical component of their survival.

There is not much literature regarding the use of fish skin by the Patagonian Indigenous Peoples, but they were known for the use of sea cow hides to make clothes and blankets to keep them warm and protected against the elements. The Tehuelche women spread them with chewed liver and then tan them by hand, rubbing them vigorously.

This paper evaluates the traditional fish skin tanning process used by agronomist Gabriel Fabian Trachter, this is a project promoted by the Secretary of Fisheries of the Province of Chubut in Patagonia, using a process easily adaptable to any location. The tanning was done with Mimosa extract and without any machines, allowing the development of fish skin tanning in areas with electricity deficit. The method does not intend to discard the use of drums, vats or drying tanks, but to show that it is also possible to tan without them. The results were tested at Ars Tinctoria laboratory to identify the potential of this traditional tanning process with a very low environmental impact.

Keywords: Fish skin, Traditional tanning, Patagonian Indigenous Peoples, Antarctica.

The use of hides and skins by hunter-gatherer societies

The use of fish skin for the construction of garments is an ancient tradition shared by Arctic societies along rivers and coasts. The specific Arctic groups with historical evidence of fish leather production are the Alutiiq, Yup'ik and Athabascan of Alaska; the Nivkh, Nanai and Ulchi Siberian peoples; the Ainu from Hokkaido Island, Japan and Sakhalin Island, Russia; the Hezhen from northeast China; and Icelanders (Palomino, 2021).

The use of hides and skins by Patagonian Indigenous hunter-gatherer societies has been widely corroborated based on ethnographic information and archaeological records (Parmigiani, 2014). The study of leather technology, one of the practices that distinguished the Indigenous peoples of Patagonia, can provide valuable information to deepen the interpretation of their culture (Marchione, 2013) and contributing to contemporary vegetable tanning techniques for leather. With the objective of obtaining information on the leather technology shared and transmitted by the Indigenous populations that inhabited the Patagonia region, data obtained from ethnohistorical and ethnographic sources on these hunter-gatherer cultures are explored in this paper.

Patagonia, including Tierra del Fuego, is a huge territory, dominated in the west and south by the rugged Andean Mountain chain, and in the east by dissected plateaux giving way to low plains (figure 1). The continuous marine waterway of the Stretch of Magallanes separates Patagonia from Isla Grande de Tierra del Fuego, and the Canal Beagle cuts Tierra del Fuego from the outer islands (McCulloch, 1997).

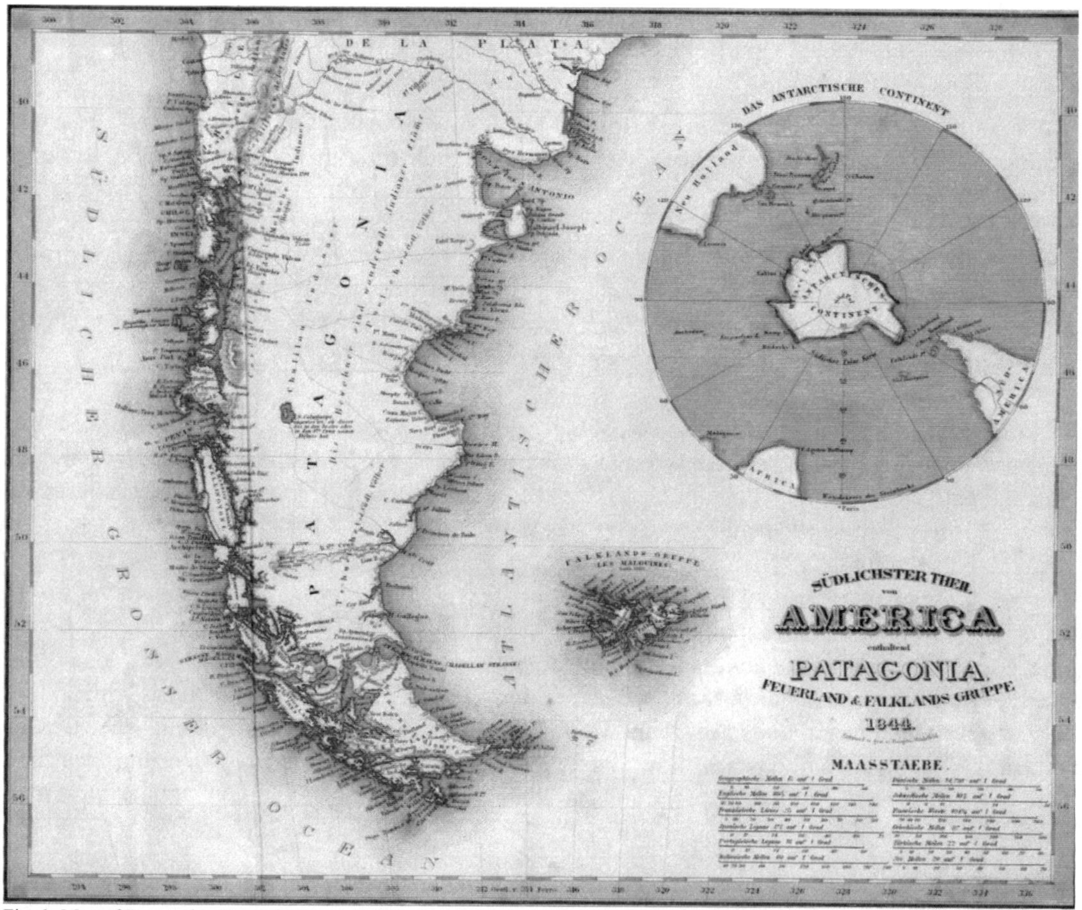

Fig. 1: Map of Patagonia, Tierra del Fuego, Falklands and Antarctic Continent. Radefeld, Carl Christian Franz, 1788-1874. David Rumsey Historical Map Collection.

The settlements in Patagonia involving the exploitation of marine resources date back to at least the 6th millennium BCE (Emperaire, 1963). The diet of these hunter-gatherer-fisherman groups was based on the marine resources available (pinnipeds, fish and seabirds) and land resources such Guanaco (Lama guanicoe) and Nandu (Rhea americana). These groups developed navigation and hunting technology at sea that allowed for a very efficient capture, as well as the systematic use of all the raw materials provided by these preys, such bones to make tools for tanning and the use of hides and skins for clothing and footwear (Orquera, 1999).

The tradition of skin and hide tanning in Patagonia combines ancestral traditions scattered in different geographical areas. The Tierra del Fuego region was inhabited by groups of hunter-gatherer-fishermen who navigated skilfully in canoes throughout the islands. These groups were the Alakaluf (figure 2), who occupied the western sector of the Beagle Channel and the southern shores of the Strait of Magellan, and the Yámanas, who inhabited the central and eastern sector of the Beagle Channel and the islands extending south to Cape Horn (Parmigiani, 2014). They lived in small family groups with camps close to the coast exploiting the resources provided by the environment, including fishing. The Selk'nam living in the central and northern part of the Isla Grande of Tierra del Fuego were not seafarers or deep-sea hunters, however, they valued sea cow hides, which they used especially for making strong straps to carry cargo (Gallardo, 1910).

The ethnographic information about the Tierra del Fuego hunter-gatherer-fishermen, shows the importance of the exploitation of marine mammals. Regarding the use of hides, it has been documented that the fur of the seal (Arctocephalus australis) was used to make various items and for exchange with other hunter-gatherers' groups (Parmigiani, 2014). The 16th century chronicles from travellers mention the use of the seals' skins for clothing, canoes and huts. F. Cortés Ojeda, in 1558, mentions an encounter with a group of men whose clothes were made with sea cow hides (Gusinde, 1974). Written sources on the Tierra del Fuego canoeists refer to the variety of skins that were used to make different goods, including clothing, footwear, household goods, bags, containers, and ropes. Many of them refer in particular to pinniped skins, which have even been used for body coverings in funerary practices (Emperaire, 1963). After contact with European navigators, the skins and artefacts made from them were taken to Europe to be part of museum collections and to inform about the diversity of Tierra del Fuego's fauna (Parmigiani, 2014).

Fig 2: Alakaluf family in their canoe 1871.

The Tehuelches, the tribes of the Patagonia area stretching from the Strait of Magellan to the Negro River, clothed themselves with short-sleeved shirts, made of overlapping sea cowhides and guanaco hides (figure 3). This garment was generally tight at the waist and in the combats, they acted as armour (Guinnard, 2007).

There is little information available on how marine mammal skins were processed, apart from few ethnohistorical descriptions of the techniques used by Tierra del Fuego canoeists. The information indicates that in the past men would oversee the hunting, skinning and cutting up the animals but the depositaries of all the knowledge linked to the tanning, scraping, oiling and sewing the artefacts were women, and this knowledge was passed on amongst women and from one generation to the next (Gómez Otero, 1996).

Ethnographic sources provide data on the use of different elements for tanning the hides: fat, raw liver, liver cooked with salt, and/or with alum dissolved in water (Musters, 2005). The Tehuelche women in charge of the manufacture of leather blankets made from sea cowhides spread them with chewed liver and they tan them by rubbing them vigorously (Guinnard, 2007).

Fig 3: Tehuelche chief wearing a Guanaco leather blanket

Fig 4: Alakaluf Fuegians, dressed in guanaco skins 1882

Patagonian fish skin tanning process

There is not much literature regarding the use of fish skin by the Patagonian Indigenous Peoples, but this paper presents the reconstruction of a leather technology tanning process from the central-northern Patagonian region. The project aims the rediscovery, preservation, reproduction, and enhancement of ancestral techniques, respectful of local traditions.

The aim of the research was to identify the biochemical logic of the traditional Patagonian tanning process using natural principles with a very low environmental impact in order to introduce them in a near future to contemporary industrial tanning, reducing the supply of chemicals, the environmental impact, generating cleaner water and less CO_2. To evaluate the process, samples of salmon leather were undertaken using skin processing technology based on traditional knowledge from Patagonian communities.

The Patagonian fish skin tanning process is part of a community-based project grounded in the desire to create greater access to Argentinian fish skin knowledge. The project is led by Gabriel Fabián Trachter (figure 5), an agronomist engineer with vast knowledge on fish skin tanning techniques. Since 2016 he has developed and taught many courses on fish skin removal, conservation and tanning techniques in National Universities and job training programmes. He has published a large amount of bibliographic material on the tanning of non-traditional skins and has also worked as manager and technical consultant of several industrial tanneries.

The treatment of the fish skins has been done with natural processes and products free of harmful chemicals, using low technology methodologies, which can be appropriated by small entrepreneurs, with an optimal use of inputs such as water, energy, and chemicals, and minimising the amount of waste generated in the tanning process, all of which are pillars of a cleaner production. The project promotes a responsible sustainable management of water and renewable resources, with a commitment to climate change. The project has been developing and rescuing steadily traditional Patagonian tanning and dyeing processes facilitating the creation of new knowledge about the fish skin craft and connecting across the Patagonian communities. The tanning and dyeing techniques of the skins has been carried out according to the traditional techniques of the inhabitants of Patagonia Argentina using autochthonous plants for tanning and dyeing as well as using Mimosa, a commercial source of vegetable tanning agent from Brazil. The project has been developed at Rawson, the capital of the Argentine province of Chubut, in Patagonia and has been directed to local communities from disadvantaged backgrounds from the remote rural areas of Rada Tilly, Sarmiento, Camarones, Rawson, Trelew and Puerto Madryn in in the province of Chubut, Patagonia.

Fig 5 to 7: Gabriel Fabián Trachter, an agronomist engineer Secretaria de Pesca de la provincia de Chubut tanning the salmon skins

The process of the fish skin traditional tanning is complicated and relatively difficult to accomplish without specialized training. The risk of disappearance of this local craft is also linked with overfishing and water pollution. Patagonian local people had applied sustainable practices in fishing for years but discarding this principle in modern times has led to the depletion of certain species. Climate change has also impacted fishing in the area. As this natural resource is reduced, the skills and technology connected to it are in increased danger of being lost (Palomino, 2021).

The project is guided by:
-Ecologically sustainable procedures, using tanning products derived from the local flora.
-Socially sustainable values, revitalising both artisanal production methods and respecting the customs of the Patagonian region.
-Economically sustainable practices, promoting the connection between artisanal manufacturing and semi-industrialised techniques.

The Patagonian fish skin project urges communities to reflect on current overconsumption and waste, to recuperate and upcycle. The aim is to learn a traditional craft with low-tech approaches and to produce a new material using the discarded fish skins from their daily diet and tanning materials available from local flora, creating a reciprocity between craftsmanship and innovation.

Fish skin tanning practices have the potential to offer important opportunities in (re)connecting Patagonian communities with common cultural heritage.

Mimosa tanning process

The salmon skins were tanned with mimosa. The mimosa tannin is obtained from the bark of a tree (Acacia mearsnii), it is a polyphenol, of small particle size, which facilitates the penetration inside the leather that is being tanned, avoiding that it fixes superficially, which would give if this happened, a hardened leather and without being tanned in its interior. The colour it gives to the skins, allows them to be dyed in a wide range of shades, which even without having the same affinity with the various dyes that Chromium has, gives results that can hardly be obtained with some other sources of vegetable tanning agents.

An interesting aspect of the wood that provides this tanning extract is that in its bark, from which the commercial product called Mimosa is extracted, there is between 20 to 30% of tannin, with a large amount of active OH groups in this polyphenol, providing a good affinity to combine with the collagen of the fish skin.

The commercial use of this Acacia can be done very fast, within seven to ten years after planting, and the heartwood can also be used for other purposes. As it belongs to the Leguminosae family, it can also incorporate atmospheric nitrogen into the soil through the association of diastrophic bacteria (rhizobia) with its roots.

As with all tannins, the pH of fixation and penetration into the raw skin should be acidic, and it is better to use weak acids such as formic or lactic acid. In addition to the supply of these provided by the chemical synthesis industries, this project is advancing in the possibilities of processing in the same establishment where the hides are processed, through simple microbiological techniques (figure 8). The microbiological pickling of the fish skins was carried out using acidification with bacteria instead of lactic acid. To reach the desired pH, the water was acidified with ferments made with bacteria. These microorganisms started to develop lactic acid and a pH4 was reached with the activity of the bacteria.

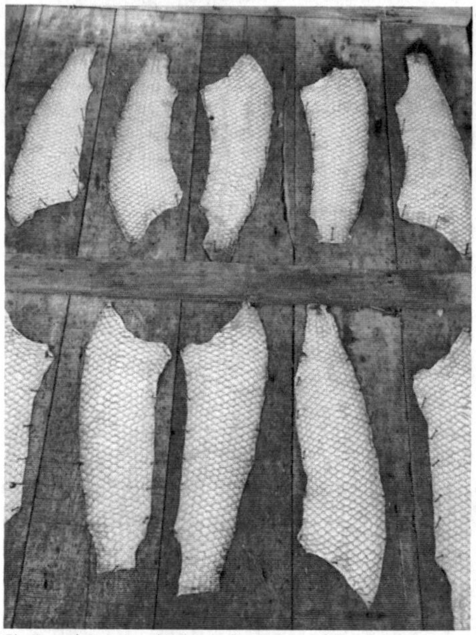
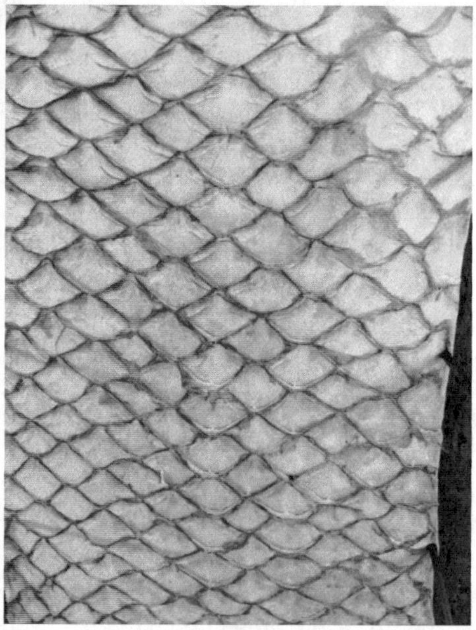

Fig 8 and 9: Microbiological pickling of the fish skins

The use of mimosa extract to tan the fish skins had the objectives of transforming the skin into leather by the formation of chemical bonds between the tannin and the collagen and the filling of the skins to improve their qualities. Using basic infrastructure and the appropriate chemical products, salmon skins suitable for any manufacturing industry were obtained. By knowing the tanning techniques, how salmon skins react to them and taking into consideration not to produce negative impacts on the environment, good results can be obtained. Likewise, if the local tanners are committed with the work they are doing, they can do it without the usual machines, which allows this activity to be carried out in areas with electricity deficit. The method does not intend to discard the use of drums, vats or drying tanks, but to show that it is also possible to tan, under certain conditions without them.

The tanning included comparative fatliquoring tests, to explain the reasons for some common defects in these skins and to show some particularities of salmon, which differentiate them from other fish species in the processes.

The fish leather was dyed without using drums, but with a lot of hand movement and rubbing the skins, so the fibres of the skins are loosened, and, in this way, there are no stains produced by sectors that react differentially with the dye. The oiling was done in a bath, with agitation during 40 min. and fixation with formic acid for 15 min. The oils used were soybean based and a mixture of soybean and fish oil.

The comparison for this type of skins between the fatliquor with soy lecithin and the mineral fatliquor is largely in favour of the former in terms of softness to the touch, flexibility, and fullness, it is also possible to achieve these results by applying this fish oil or another of animal origin manually on the underside and possibly on the grain side, thus giving other characteristics. From previous experience, oils of animal origin also have excellent results.

All the processes of the wet phase and tanning processes, were carried out by immersion with manual movement, thus replacing the action of the drums.

The tanned salmon skins have the particularity of providing a good shine when polished. The shine finished of the skins (figure 11) is the result of the friction given using a polishing machine (figure 10) with a mechanical arm and a cylindrical agate stone. If this tool is not available, it can be done by hand with smooth stones or glass bottles. Salmon has a good reaction to polishing, without the need to apply finishing products, such as modified caseins. In this case, no casein or other finishing products were used.

Fig 10: Polishing machine with a mechanical arm and a cylindrical agate stone

Fig 11: Polished salmon skins

Fisheries waste in Patagonia

Fishery waste has increasingly been incorporated into the value chain of fishery products in countries like Iceland, generating new lines of industrial production of high value-added products. Currently, in various parts of the world, fish waste, consisting of skins, viscera, and bones, is used to produce by-products, including fish meal for animal consumption and other products such as protein concentrates, pharmaceuticals, fertilisers, glues, gelatine, and oils to tan skins (Palomino, 2020).

In Argentina, most of the companies dedicated to fish processing do not make use of the fish waste, despite the fact that the technology exists to do so on both an artisanal and industrial scale, making them a material that could be integrated as a primary or secondary input in the manufacture of clothing and accessories, and bringing together actors to integrate tanned skins from marine and inland water fish into the food value chain generating high value-added products.

The fish skins used in this project have been collected from a freshwater fishing company in Musters Lake in Sarmiento, Chubut. The skins were supplied by an artisanal commercial fisherman (figure 12) who catches rainbow trout, perch and silverside. He goes fishing daily and has several gill nets that he collects and re-sets.

Fig 12: Freshwater fishing company in Musters Lake in Sarmiento, Chubut, Patagonia.

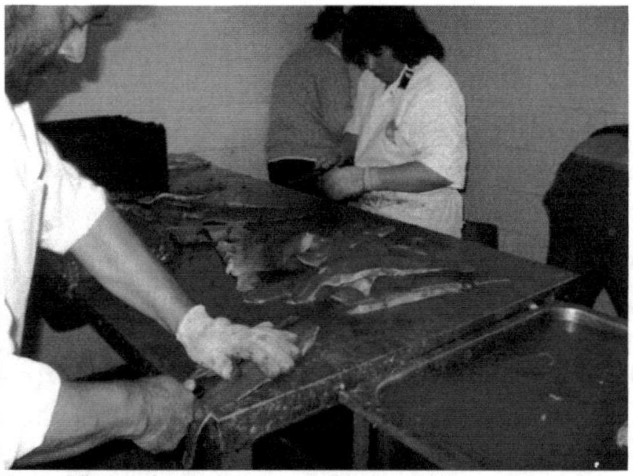

Fig 13: Filleting the rainbow trout.

Based on the recognition of the diverse geographical conditions of Argentina and its regional fish fauna, the use of different species could benefit the fish leather business. Thus, the northern region of the country could provide skins from river fish, while the eastern coastline could provide fish from brackish waters; and from the south, fish from lakes and meltwater rivers. Each type of skin could have its own palette and textures that reflect the indigenous cultural repertoire of each community.

The market study of Patagonian fish leather, carried out by the local INTI agency (National Institute of Industrial Technology) made possible to explore and measure the market need for products made from fish leather, estimating the prices at which they could be sold and detect the channels through which they could be marketed.

The improvements and simplifications in the tanning processes and the changes towards the use of organic products in tanning that have been implemented in this project, allows the adaptation of these materials to the new market trends that seek the use of raw materials with certificates of good environmental practices.

Test Results

Several tests were carried out to understand the mechanical, tensile strength and physical properties of the tanning methods, their resistance to breakage under tension or how well the skin will perform in vigorous use. By testing the results obtained with different Patagonian fish skins tanning techniques, we gained a better understanding of their physical properties and their limitations. The mimosa tanning technique allows fish skin to becoming a truly all-purpose material, able to be adopted for its practical as well as aesthetic capacities. The tensile strength of fish skins was tested in a dynamometer considering the acceptable values needed for different applications, sampling probes parallel and perpendicularly to backbone.

Report AN3861-21/1 Re-oiled vegetable tanned leather - dyed brown.
The resistance to tearing was good, considering the low leather thickness (0.41 mm). Normally, the minimum acceptable value of resistance to tearing is 20 N. The leather has a good resistance to tearing.

Report AN03860-21/1 Salmon leather 1 - Mimosa tanned, brown dyed, polished with agate stone.
It has much better tear resistance (51N), almost twice as much as the previous one. The media thickness measured was 0,57 mm, giving a specific tear resistance value obtained reached 96,5 N/mm, which is an interestingly high value.
Regarding Martindale abrasion, the result was low due to the scales pockets (like snakes' skins).

TEST N. 2 - Leather — Physical and mechanical tests — Determination of tear load — Part 2: Double edge tear

OPERATING CONDITIONS: (23±2) °C , (50±5)% R.H. - N°of tested specimens: 3 parallel to the backbone, 3 perpendicular to the backbone

INSTRUMENT: Electronic dynamometer

PARAMETER	*RESULT*	*MEASURING UNIT*
medium thickness - parallel to the backbone	**0,57**	mm
medium thickness - perpendicular to the backbone	**0,52**	mm
arithmetic mean of medium thickness - parallel to the backbone / perpendicular to the backbone	**0,55**	mm
Average load at the parallel tear to the backbone	**42**	N
Average load at perpendicular tear to the backbone	**59**	N
Arithmetic average of the average load - parallel / perpendicular	**51**	N

Fig 13: Physical and mechanical tests. Determination of tear load.

The tensile strength of fish skins was tested applying sampling parallel to backbone and perpendicularly. The analysis of the results shows that the smallest marginal stresses occur along the backbone of fish skins and the smallest marginal deformations occur perpendicularly. This distribution of strength properties differs from the distribution of strength properties in bovine hides. The orientation of collagen fibres in a linear direction is adapted to the aquatic environment of the fish, therefore fish leather is found to be more stretchable along its length than width (Palomino, 2021).

From the results of the tests carried out on these salmon leathers, it is clear that the tear resistance is very high considering mammals leathers of that thickness, while when polished with a machine with mechanical arm and stone, this value increases almost twice as much, possibly as a consequence of the crushing of the fibres, product of the mechanical action carried out by the machine, which also produces a better dispersion of the oil incorporated in the leathers, in the oiling process. The excellent tear resistance observed is due to the peculiar fibre's orientation typical of fish skins. In other projects fish skins were already used in the footwear and leather goods industries, among others, with excellent results in terms of resistance to traction and to tearing.

Conclusions

Skin processing technology is based on traditional knowledge of materials: the knowledge to physically manipulate materials to acquire the properties that are needed for a specific purpose, and the ability to adapt materials and methods according to current conditions (Klokkernes,2007). Through this research we have identified the biochemical logic of the Patagonian traditional fish skin tanning process using natural principles with a low environmental impact. These traditional, environmentally friendly techniques can be used in combination with modern technology reducing the supply of chemicals, generating cleaner water, less CO_2 and minimising the environmental impact of current fish leather production.

The vegetable tanning of fish leather is time-consuming compared to industrial tanning and needs a different set of skills. However, it has unique properties. Regarding the hypothesis about the possibility of continuity in the practice of leather working between pre-Hispanic populations and ethnographic and ethnohistoric groups, the analysis carried out provides elements to sustain that the technical knowledge - the "know-how" - has remained in place until modern times. This work has allowed us to delve into an underdeveloped subject in reference to leather technology and its approach from ethnohistory. This proposal and the methodology used can be improved and expanded, which will enable a much richer and deeper approach to the ways of life in Patagonia's past. This paper aims to document Patagonian fish skin knowledge systems and practices which are in danger of disappearing. The paper shares fish skin skills and tanning practices that remote Patagonian communities are currently developing.

The project aims to strengthen knowledge networks between Patagonian craftspeople, disadvantaged communities and local artists. The project hopes to preserve the fish skin craft and contribute to sustainability practices through the study of the material and the transmission of fish skin tanning skills.

The fish skin manufacturing process has brought to light also finished objects such mate holders, wallets, and shoes. The adaptation of the production processes for their subsequent manufacture in the place of origin of the fish waste raw material has increased the development of local industry with the added value of aesthetic traces that carry their designation of origin so local communities can replicate them for their further commercialisation.

Acknowledgements

This research has been funded by the Secretaria de Pesca de la provincia de Chubut, Argentina and the European Commission Horizon 2020 research and innovation initiative under the Marie Sklodowska Curie agreement FishSkin 823943. Our biggest gratitude goes to Gabriel Fabián Trachter for his continuous work developing sustainable fish skin tanning processes. All the fish skins used in the project have been tanned by him. We would like to thank Gustavo Defeo and his Ars Tinctoria laboratory where the fastness, physical and mechanical properties tests were performed and to Patricia Casey, past president of the I.U.L.T.C.S. In addition, this research could not have been completed without the support from William Fitzhugh, director of the Arctic Studies Center at the National Museum of Natural History, Smithsonian Institution and his team: Stephen Loring, John Cloud, Nancy Shorey, Aron Crowel, Dawn Biddison, and the Fulbright UK US scholar award, funding the Arctic Indigenous fish skin research of Elisa Palomino.

Bibliography

Chapman, A. (1986). Los Selk'nam. La vida de los Onas. Buenos Aires. Emecé.

Cook, J. (1777). A voyage towards the South Pole and round the World, performed in His Majesty´s ships the Resolution and Adventure in the years 1772-1775. London

De Angelis, H. H. (2015). Arqueología de los cazadores-recolectores de la faja central de la Isla Grande de Tierra del Fuego. Bs As. Sociedad Argentina de Antropología.

Emperaire, J. (1963). Los Nómades del Mar. Ediciones de la Universidad de Chile.

Gómez Otero, J. (1996) Video "Quillanguería", sin publicar, realizado en el marco del proyecto "Rescate y reinserción de la quillanguería patagónica en el ámbito de la provincia del Chubut", Fundación Trafyén-Encontrarse. Centro Nacional Patagónico (CENPAT).

Guinnard, A. (2005) Three Years Slavery Among the Patagonians. Continente. Nonsuch Publishing

Gallardo, C. (1910). Tierra del Fuego: Los Onas. Cabaut.

Gusinde, M. (1982). Los indios de tierra del fuego: los Selk`nam. Centro Argentino de Etnología Americana. Buenos Aires.

Gusinde, M. (1986). Los indios de Tierra del Fuego: los Yamana. Centro Argentino de Etnología Americana. Buenos Aires.

Gusinde, M. (1974). Los indios de Tierra del Fuego: los Halakwulup. Centro Argentino de Etnología Americana. Buenos Aires.

Klokkernes, T. (2007) Skin processing technology in Eurasian reindeer cultures. The Royal Danish Academy of Fine Arts the School of Conservation.

Mansur M. E., Parmigiani, V. V. (2014). Pieles y cueros. Cadenas operativas en la producción y uso de bienes por los pueblos originarios de Tierra del Fuego. Comunicación presentada en las IX Jornadas de Arqueología de la Patagonia, Coyhaique.

Marchione, P.C., Bellelli, C. (2013) El trabajo del cuero entre los cazadores-recolectores de la Patagonia centro-septentrional. Campo Moncada 2(Valle medio del rio Chubut). Relaciones de la Sociedad Argentina de Antropología XXXVIII (1).

McCulloch, R. D., Clapperton, C. M., Rabassa, J., & Currant, A. P. (1997) The Natural Setting: The Glacial and Post-Glacial Environmental History of Fuego-Patagonia. In C. McEwan et al (Eds.) Patagonia: Natural History, Prehistory and Ethnography at the Uttermost End of the Earth (pp. 12–31). Princeton University Press.

Musters, G. Ch. (2005) Vida entre los Patagones. Un año de excursiones por tierras no frecuentadas desde el estrecho de Magallanes hasta el Río Negro. Buenos Aires, El Elefante Blanco.

Orquera, L. Piana, E.L. (1999). La vida material y social de los Yámana. Eudeba-IFIC, Buenos Aires.

Palomino, E. Rahme, L. (2021) Indigenous Arctic Fish Skin-A study of different traditional skin processing technology. Society of Leather Technologists and Chemists Journal. Vol. 105 issue 2.

Palomino, E. (2020) SDG 14 Life Below Water. Introducing Fish Skin as a Sustainable Raw Material for Fashion In: Franco I., Chatterji T., Derbyshire E., Tracey J. (eds) Actioning the Global Goals for Local Impact. Science for Sustainable Societies. Springer, Singapore. ISBN: 978-981-32-9927-6.

Palomino, E. Defeo, G. (2019) Material Design Research - Fish skin, a new environmental -friendly material for fashion. Design Research for Change. Design Museum, London ISBN 9781862203693.

Parmigiani, V., Alvarez Soncini, M. C., Mansur, M. E. y Martinoli, M.P. (2014) Processing of fur seal leather (Arctocephalus australis) between southern maritime groups: an experimental evaluation.

Tivoli, A. M., Zangrando, A. F. (2011) Subsistence variations and landscape use among maritime hunter-gatherers. A zooarchaeological analysis from the Beagle Channel (Tierra del Fuego, Argentina). Journal of Archaeological Science, 38(5).

Formatting Issue - Best Available Paper/Graphic

Preparation and application of polyether amine triazine derivatives as a chrome-free tanning agent for cleaner production

XinjuJia [a], Ran Tan[a], Biyu Peng [a, b, *]

[a] National Engineering Laboratory for Clean Technology of Leather Manufacture, Sichuan University, Chengdu 610065, PR China

[b] Key Lab. of Leather Chemistry and Engineering of Ministry of Education, Sichuan University, Chengdu, Sichuan 610065, PR China

ABSTRACT: In this work, a polyether amine triazine derivatives (ET) was prepared via one-step process and its structure was confirmed using FT-IR, 1H NMR analyses. Then, the ET was applied to the tanning process and a novel chrome-free tanning approach has been optimized and evaluated. The result indicated that after tanning with 8% of ET at pH6 for 6h, the shrinkage temperature of leather was 83.9℃ and the thickness change rate was 97.5%. The ET tanned leather was endowed with a close mechanical and organoleptic performance as conventional chrome tanned leather. The XRD, DSC, TG, SEM and Zeta potential analyses demonstrated that ET could couple with the amino group of collagen side chain and breaking partial hydrogen bond between fibers, brought significantly improvement to the softness of leather, but has no impact on the high-level structure of collagen. Another promising features of this tanning approach is that the residual Cl- concentration in wastewater decreased 42.1% in tanning process compared to conventional chrome tanning method. The emissions of COD and TDS in tanning process reduced by 40.5% and 27.2%, respectively. It is conceivable that the ET tanning approach satisfactorily solves the longstanding problems of conventional chrome tanning and provides an improved approach to promote leather industry to develop sustainably.

Key words: Polyether amine; Chrome-freetanning; Environmental friendly; Cyanuric chloride; Triazine derivatives

Formatting Issue - Best Available Paper/Graphic

* Corresponding Author e-mail: pengbiyu@scu.edu.cn; Tel.+86-28-85401208.

1. INTRODUCTION

As the most critical step in leather manufacturing,tanning is a process in which chemicals are used to introduce stable cross-linking between collagen and convert raw hide into leather.Chrome tanning method is the most popular tanning method in the leather industry due to the prominent tanning performance of chrome.However, because of the low uptake of chrome in tanning process, about 170,000 tons of Cr wastes/year is released into environment worldwide[1]which could lead to consume of resources and environmental pressure.In recent years, with theestablishment ofstrategies of sustainable development, extensive studies have been conducted todevelopsubstitute tanning agents of chrome.Aldehyde derivatives (such as modified glutaraldehyde[2],oxazolidine[3], organic phosphorus[4] have received special attention among researchers due to the significant potential in protein cross-linking.The aldehyde derivatives tanning couldendowed wet-white with excellent hydrothermal stability and physical properties[5], which has been regarded as an appropriate alternative of conventional chrome tanning.Although an eco-friendly tanning technology seems can beachieved through aldehyde derivatives, there are somespecific problems can still be encountered.The principle of aldehyde derivatives tanning is basically the aldehyde group or hydroxymethyl groupcoupling with the amino group of collagen side chain [6]. But thecrackingof hydroxymethyl group and present of aldehyde group could lead to the formaldehyde emission of wet-whiteduring storage[7, 8].Because of the biological toxicity of formaldehyde,it has been classified as a potential human carcinogen by the International Agency for Research on Cancer (IARC) and the products contain formaldehyde were facing stringent environmental regulations[9].To overcome the aforementioned limitations, a novel tanning agent is needed to be employed without chrome or aldehyde and have good tanning property to meet the needs of the market.

Cyanuric chloride(TCT) is an important chemical intermediatewhich have many applications in synthesis of reactive dyes[10, 11]. The chlorine group in TCT have strong reactivity with collagen under the action of triazine ring.Specifically, thethree chlorides can be substituted with the amino groups of collagen side chain at 0-5℃, 35-40℃, 75-80℃, respectively[12, 13]. When the TCT react with protein, the optimal pH for the reaction of the second Cl- on the TCT with the collagenis 9.4 and the main reaction site is the amino group on the lysine residue[14], significant improvement has been found in the thermal stability and physical and mechanical properties of the protein treated with TCT[15, 16].Besides, the rigid structural such as triazine ring of TCT could also bring stability of crosslinking[17].However, the poor water solubility of TCT makes it unachievable to direct apply in aqueous phase tanning, The commonly used methods is to couple TCT with sulfonated aromatic compounds to enhance the water solubility [15], but the introduction of aromatic compounds could increase the molecular size of tanning agent and prevent the penetration of tanning agent in skin. Besides, the active chlorine content of commercial triazine derivatives tanning agents is limited considering that the third chlorine in the structure of cyanuric chloride could be difficult to be involved in the crosslinking reaction, which leads to low crosslinking rate and the shrinkage temperature[18].

In this work, anovel chrome-free tanning agent based on polyether amine triazine derivatives (ET)was synthesized.The introduction of the polyether amine as backbone endowed the product with good water solubilitywithout excessive increase in molecular size and there aremulti-site of active chlorine in product structure.The structure of ET was characterized and theeffect of ET alone use in the main tanning was investigated andevaluatedby comparing with the commercial triazine derivatives tanning agent (CTDT) and the chrome tanning agent.Based on the experimental results,a novel chrome-free aldehyde-free approach for eco-friendly wet-white manufacturing was developed.

2. EXPERIMENTAL

2.1 Materialsand Equipment

The raw material of cowhide was used for leather processing. Polyetheramine (ED) was purchased from Suzhou ChangkeNew Materials Technology Co., Ltd. Cyanuric chloride was purchased from Aladdin Reagent Co., Ltd.The chemicals employed in the subsequent operations were those normally used in leather industry.

Stainless Experimental Drum (GSD400-4, Wuxi Xinda Light Industry Machinery CO., Ltd).The digital leather shrinkage temperature tester(MSW-YD4) was produced by Shaanxi University of Science & Technology, China.

2.2 Preparation of polyether amine triazine derivatives (ET)

0.1 mol of cyanuric chloride and 100mL of THF (dehydrated) and 0.25 mol of sodium carbonate (dehydrated) was chargedin a 300 ml 3-necked flask equipped with stirring bar. 0.0245 mol ofED (dehydrated) was add dropwise within 0.5 h under the ice bath. After the dropping of ED, the reaction took placed at 25°Cfor 4 h under the high speed stirrer. The viscous liquid ET was obtained by removing THF through rotary evaporation.

2.3 Fourier Transform Infrared Spectroscopy (FT-IR)

The functional groups present in the ET were characterized using the FT-IR. The FT-IR was used to determine the pure sample at room temperature by using Nicolet iS10 infrared spectrometer. The sample was grinded and pressed by spectral pure KBr, and scanned 32 times in the wavenumber range of 400 to 4000 cm^{-1}. The resolution was 2 cm^{-1}.

2.4 Nuclear Magnetic Resonance spectroscopy (^1H NMR)

After purification and drying, the product was analyzed by 400MHz Brucker ARX400 nuclear magnetic resonance spectrometer. The solvent was deuterated chloroform $CDCl_3$ and the internal standard was tetramethylsilane TMS.

2.5 Leather tanning processing

The tanning process were carried out using depilated and softened cattle hide,the ratio of leather chemicals of tanning process was calculated according to the limed split pelt weight and the cumulative input of tanning process was presented in **Table. 1**.

Table. 1 ET tanning process

Processes	Chemical	Dosage/%	Temperature/°C	Time/min	Remarks
Wash	water	400	25	20	Repeat twice
Tanning	water	50			
	ET	X		30	Initial pH 6.0
Basification	Na_2CO_3	1.5		180	DesignatedpH Y

	water	150	Z	150	overnight
Horse up					

X: The dosage of ET depending on the experiment:4%, 8%, 16%; Y: ThepH depending on the experiment: 3-9; Z: TheTemperature depending on the experiment: 25℃35℃, 45℃

The control process was carried out using commercial triazine derivatives tanning agent (CTDT) and the cumulative input of tanning process was presented in **Table. 2**

Table. 2CTDT tanning process

Processes	Chemical	Dosage/%	Temperature/℃	Time/min	Remarks
Wash	water	400	25	20	Repeat twice
Tanning	water	50			
	CTDT	9		240	
	water	30	50	240	pH5
	Fungicide	0.1		60	overnight
Horse up					

The subsequent neutralization and fatliquoring process was carried out according to the conventional process.

2.6 Determination of the shrinkage temperature

The similar parts were selected for sampling and the leather samples was measured byleather shrinkage temperature tester (MSW-YD4 Sunshine ElectronicResearch Institute of Shaanxi University of Science and Technology).Each experimental plot wasobtained from an average of three samples.

2.7 Determination of the exhaustion rateof tanning agent

The waste liquor before and after tanning was collected and filtered,then digested with acid at 150℃ for 2hrs.The digestion solutions were appropriately diluted the chemical oxygen demand (COD) were analyzed by using the DR1010 COD detector. The a_1 (ppm) stand for COD in waste liquid before tanning while the a_2 (ppm) stand for COD in waste liquid after tanning. The exhaustion rate of tanning agent is calculated according to the following formula:

$$\text{exhaustion rate}(\%)=(a_1-a_2)/a_1\times 100\%$$

2.8 Determination of the thickening rate

Select four corners and five different parts in the middle of the leather, measure the thickness with thickness meter, get the average thickness by taking the arithmetic mean. The d_2 (mm) stand for the thickness after tanning while the d_1 (mm) stand for the thickness before tanning. The δ (thickening rate) can be calculated according to the following formula:

$$\delta\ (\%)=(d_2-d_1)/d_1\times 100\%$$

2.9 Thermal analyses

Thermogravimetric analyses (TG)measurementswere carried out on the tanned leathers, the untanned hide was used as control simple. The dried samples were put intoceramic crucibles and heated with 10°C/minheating rate in a N_2atmosphere. The range of temperature wasfrom 40 to 800°C A NETZSCH TG 209 F1 thermal gravimetricanalyzer (Germany) was used for the determination.

Differential scanning calorimetry (DSC) measurementswere carried out on the tanned leathers, the untanned hide was used as control simple. The dried samples were put intoaluminum crucibles and heated with 10°C/min heating rate in aN_2atmosphere. The range of temperaturewas from 0 to 200°C A DSC 200 PC differential scanning calorimeter (Germany) was used for the determination.

2.10 Zetapotentialanalyses

The leather was cut into small pieces, freeze-dried and crushed to powder, then dissolved in deionized water. The isoelectric point was determined by SZP-10 potentiometric analyzer.

2.11 Morphological analyses

The morphologies of crosslinked collagenfibersand grain patterns were observed using Desktop Scanning Electron Microscope. Samplesof each tanning groups were freeze-dried,and then observed with an accelerating voltage of 5 kV.

2.12 Whiteness analysis

The Lab value of tanned leather was determined by color reader CR-13 Portable Colorimeter and compared with the standard reference. The total color difference ΔE was calculated according to the following formula

$$\Delta E=\sqrt{\Delta L^2 + \Delta a^2 + \Delta b^2}$$

2.13 X-ray diffraction (XRD) measurement

The sample was analyzed by EMPYREAN X-ray diffractometer, the untanned hide was used as control simple. Cuk as the ray source, the voltage of the tube was 40kV, the current of the tube was 30mA and the scanning speed was 1°/ min, angle range 5-55°.

2.14 Determination of physical and mechanical properties

The leather from different groups were kept under standard atmospheric conditions for 48h prior to testing. The physical and mechanical properties i.e. the tensile strength, elongation at break, tear strength as well as the bursting strength of the resultant leather were determined. These tests were performed according to the standards of IUP 6,2000; IUP 8, 2000; IUP 9, 1996.

2.15 Environmental impact assessment

The chemical oxygen demand (COD) were analyzed by using the standard procedure(AWWA, 1998) for the spent tanningliquors. Total dissolved chlorides (TDC)wasdetermined using Thermo orion 9617BNWP chloride ion selective electrode. The total dissolved solids (TDS) was calculated after one hour desiccation at 104°C.The content of chromium in spent liquor was determinedusing Inductively Coupled Plasma Emission Spectrometer (AES-ICP)after digested.

3. RESULTS AND DISCUSSIONS

3.1 Characterization of ET

FT-IR analyses:Fig. 1 shows the FT-IR spectrum ofTCT and ET.It can be seen from the figure that the TCT and ET showed obvious stretching vibration absorption of C-Cl bond at 532 cm^{-1}, while the characteristic absorption peak of triazine ring appears at 1303 cm^{-1},1350 cm^{-1} and 1546 cm^{-1}. The C-N characteristic peak of ET at 1103 cm^{-1}could not be observed in structure of TCT which indicates that the polyether amine was introduced into the structure of the product.

Fig. 1 FT-IR spectrum of ET

^{1}H NMR analyses: Fig.2 shows the ^{1}H-NMR spectra of ET while using TMS as internal standard and deuterated chloroform as solvent.As can be seen from the figure, the chemical shift at 7.26 ppm belongs to the deuterated chloroform, the chemical shift at 1.81 ppm was caused by the residual solvent of tetrahydrofuran, thechemical shift at 3.70 ppm was caused by the ethyl hydrogen in the structure of polyether amine. It can be seen that the characteristic hydrogen in the structure of tanning agent can be assigned, which proves the existence of ET.

Formatting Issue - Best Available Paper/Graphic

Fig. 2 ^{1}H NMR spectrum of ET (CCl_3D, 400MHz)

3.2 Optimization of the ET tanning conditions

Optimization of the ET tanning pH: In order to prevent the excessive hydrolysis of tanning agent at the initial stage of tanning, the pH of tanning system was controlled at 6 at the initial stage of tanning,8% of ET was used for tanning and the sodium carbonate solution was used to adjust the pH at varying pH, the thickening rate and shrinkage temperature of wet-white was investigated.**Fig.3** shows the thickening rate and shrinkage temperature of wet-white after tanning at different pH.It can be seen from the figure that during the tanning process, the shrinkage temperature and thickening rate of wet-white slightly increased as the pH increased from 3.0 to 6.0 and rapidly decreased at higher pH.The shrinkage temperature reached the maximum of 79.8 ℃ and the thickening rate reached the maximum of 86% at the pH 6. This is due to the unique hydrolysis-binding equilibrium theory of TCT[12]. When pH is lower than 6, the binding ability of active chlorine to collagen is insufficient, resulting in weak tanning property, low shrinkage temperature and thickening rate; When the pH is higher than 6, high pH will lead to the rapid hydrolysis of TCT[19], resulting in the loss of tanning property and the decrease of cross-linking rate. Therefore, it is an appropriate choice to tanning at pH 6.

Formatting Issue - Best Available Paper/Graphic

Fig. 3 Shrinkage temperature and thickening rate of wet-white tanned with ET at different pH

Optimization of the ET tanningtemperature: Temperature is another important factor affecting the reaction between ET and collagen.For the leather manufacturing, a lower temperature is usually used to promote the penetration of tanning agent in the initial tanning stage, and a higher temperature is used to promote the combination of tanning agent and collagen in the final tanning stage, which makes the tanning effect uniform. In order to investigate the effect of temperature at the final tanning stage on ET tanning properties, 8% of ET was used at room temperature for 3h at the initial stage of tanning, and varying temperatures of 25°C, 35°C and 45°C were used at the final stage of tanning,the shrinkage temperature of wet-white was investigated at different times.The resultin **Fig. 4**shows that the temperature rise at the final stage of tanning had a significant promoting effect on the shrinkage temperature of wet-white.High temperature not only could accelerate the shrinkage temperature rise of wet-white, but also increase the final shrinkage temperature of wet-white to a certain extent.The shrinkage temperature of wet-white could reach 79 °C after be treated at 45 °C for 150 min. Considering that further heating may cause the destruction of collagen structure and excessive the hydrolysis of ET,it is suitable to treat at the temperature of 45 °C for 150 min at the final stage of tanning.

Formatting Issue - Best Available Paper/Graphic

Fig. 4 Shrinkage temperature of wet-white tanned with ET at different temperature

Optimization of the ET dosage: The tanning agent of 4%, 8% and 12% was used for tanning according to the process mentioned in**Table.1**. The subsequent neutralization and fatliquoring process was carried out according to the conventional process.As the **Table. 3**shown, with the increase of ET dosage, the shrinkage temperature and thickness of wet-white increased obviously. When the dosage of ETreached 8%, the shrinkage temperature of crust leather was 83.9℃ the thickening rate was87.3% and no significant increase was achieved with the further increase of ET dosage. As for the exhaustion of tanning agent, when the ET dosage was higher than 8%, the exhaustion rate of ET decreases obviously, which indicates that the combination of ET and leatherwasreaching saturation. Further increasing the ET dosage contributea little to the leather properties.Besides, a large amount of ETremained in the effluent liquor could increases the difficulty of post treatment. In order to ensure the effective utilization of materials and control the cost, it is considered that 8% of ET dosage could be a more appropriate choice.

Table.3Tanning effect of different ET dosage

ET dosage/%	4	8	12
Ts of the wet-white /℃	73.9	78.9	79.3

Ts of the crust leather /℃	76.1	83.9	84.1
thickening rate /%	57.1	87.3	90.1
exhaustion rate %	81.6	79.6	59.3
Softness	76	88	89

3.3 Thermal analysis

The thermodynamic analysis of leather is a reliable method to evaluate the thermal effect of chrome-free tanned leather. It can also reflect the effect of crosslinking on the advanced structure of leather. In this section, the thermal performance of wet-white tanned with ET was analyzed via TG and DSC, the untreated collagen was used as the control.

The pyrolysis property of leather is related to the cross-linking degree of collagen fibers. The cross-linking degree of collagen fibers is negatively related to the collagen mass of thermal decomposition[20].**Fig. 5** showed the TG curves of the ET, and the corresponding characteristic parameters of them are given in **Table. 4**. The first stage (I) is the vaporization of free water and bound water, the second stage (II) is the decomposition of collagen fibers, polymers and organic chemical materials used in tanning, and the third stage (III) is the decomposition of chemical bonds of polymers with collagen[21]. According to the results,the crosslinking of ET can obviously slow down the thermal decomposition during the heating process, this is because the triazine ring structure in et can form stable sites between the collagen chains and play a role in fixing the collagen chains, but these are amorphous structures, which have limited effect on improving the thermal stability.

The thermal stability of different leather powder before and after tanning can be judged by the position, width, strength and area of the peak in DSC spectrum.The differential scanning calorimeter (DSC) has been used to elucidate the degradation profile of wet-white tanned with ET in response to heat and the thermal stability of collagen before and after tanning can be judged by the position, width, strength and area of the peak in DSC spectrum. As showed in **Fig. 6**, thedecomposition peak of wet-white rises to 97.5℃ while the decomposition peak of untanned collagen at 69.9℃, and the position of the peak obviously drifts to the high temperature zone and get wider,That is due to the stabilizing effect of the triazine structure on collagen advanced structure.

Formatting Issue - Best Available Paper/Graphic

Fig. 5TG curve of wet-white tanned with ET

Table. 4Weight loss in different thermogravimetric stage

Simple	Thermal decomposition stages		
	I /wt%	II /wt%	III /wt%
ET tanning	8.03 (Room temperature-210.8°C)	50.11 (210.8°C493.8°C)	9.88 (493.8°C600.0°C)
Control	7.19 (Room temperature-209.2°C)	51.0 (209.2°C397.3°C)	8.63 (397.3°C600.0°C)

Fig. 6DSC curve of wet-white tanned with ET

3.4 Zeta potential analysis

Collagen is a typical amphoteric polyelectrolyte. The blocking state of -COOH and -NH_2 in its structure can be described through zeta potential to a certain extent. Specifically, when the carboxyl group is blocked, the amino group is protonated and the molecule carry positively charged; in the alkaline range, the carboxyl group is ionized, the peptide chain is hydrolyzed and broken, and the molecule carry negatively charged. When the concentration of negatively charged carboxyl group is equal to that of positively charged amino group in solution, collagen will be in the state of equal charge[22].

In this study, the Zeta potential of wet-white tanned with ET was investigated and the untreated collagen was used as the control.**Fig. 7** shows theZeta potential of wet-white tanned with ET at different pH.After tanning with ET, the isoelectric point of collagen decreased from 4.71 to 2.85, that means ET could react with the amino groups of collagen, a large number of amino groups in the side chain of collagen were blocked,which indicated that the enhancement of acid resistance of wet-white, which is consistent with the results in **Fig. 3**

Fig. 7 Zeta potential of wet-white tanned with ET

3.5 XRD analysis

In order to reveal the dispersion of ET in collagen,the white-wet tanned with ET was characterized by XRD, and the untreated collagen was used as control group. In**Fig. 8**, there were two obvious diffraction peak at 8° (A) and 22° (B). The (A) peak reflects the distance between the molecular chains of the collagen fibers, and belongs to the a-helix region with more ordered structure [23].The peak shape of (A) got sharper, which indicated that the tanning process of ET broke part of hydrogen bond and van der Waals force between collagen fibers to a certain extent, loose the fiber. The (B) peak primarily due to the diffuse scattering caused by numerous structural layers inside the collagen fibers, such as the b-fold with irregular zigzag structure [24]. After tanning with ET, the B peak moved to the direction of low diffraction angle, which indicated that crosslinking had an effect on the secondary structure of collagenand reduce the randomness of collagen braiding. There was no obvious change of the diffraction peak at 43 °,which indicated that ET tanning did not cause the change of collagen crystalline region.

Fig. 8XRD curves for wet-white tanned with ET.

3.6 Physical performances of leather

In order to evaluate the physical performances of wet-white tanned with ET, wet-white was prepared with CTDT (9%) according to the process mentioned in **Table. 2**andwet-blue was preparedwith conventional chrome tanning method (6.5%). Thesubsequent neutralization and fatliquoring process was carried out according to the conventional process and the results were showed in **Table. 5**. As the resultsshown, the physical performances of ET tanned leather were better than that of CTDT tanned leather, which was close to the physical performances of chrome tanning leather. This is because there were more active chlorine in ET structure than CTDT, which could be regarded as crosslinking sites when ET react with collagen.In addition, ET tanning can significantly improve the softness of leather, which was due to the supporting effect of triazine structure between leather fibers, and the lubricating effect of polyether chain segments in the gap between collagen fibers [25], which gives leather excellent softness.

Table.5Physical performances of leather tanned with different method

Tanning agent	ET	CTDT	Chrome
Shrinkage temperature (°C)	83.9	74.3	109.2
Tensile strength (MPa)	16.19	14.1	18.31

Formatting Issue - Best Available Paper/Graphic

Elongation (%)	65.1	66.9	65.2
Tear strength (N/mm)	47.93	38.9	60.21
Softness (mm)	8.8	9.1	7.8
Fineness of grain	++++	+++	++++
Fullness	+++++	++++	+++++

3.7 Whiteness analysis

The Lab values of leather were determined and compared with the standard reference to evaluate the whiteness of several leather tanned with different method.It can be seen from **Table.6** that the color difference between ET tanned wet-white and standard white is little, which is similar to that of CTDT tanned wet-white and obviously better than that of chrome tanned wet-blue. Besides, triazine derivatives can be coupled with subsequent dyes lead to a fixation effect, which present a wide application prospect.

Table.6 Whiteness of wet-white tanned by different methods

Tanning method	L	a	b	ΔE*
Standard white	91.60	-0.47	5.10	—
Chrome tanning	42.71	-0.18	0.88	49.07
CTDT tanning	79.30	-0.26	3.80	12.37
ET tanning	79.82	-0.49	3.77	11.85

3.8 Morphology analysis

Fig.9 showed the SEM observation of leather tanned with three different method.It can be seen from the result that the fibers of wet-white tanned with ET were more dispersed compared with the conventional chrome tanned leather, there was no obvious adhesion between fiber bundles, and the boundary of basic fibers was clear. Compared with the wet white tanned with CTDT, The looseness of the fiber was similar, which indicates that the hydrophilicity of polyether chain in the ET structure was conducive to the entry of subsequent materials.

Formatting Issue - Best Available Paper/Graphic

Fig. 9SEM observation of fiber patterns of the leather with different tanning method.

3.9 Environmental impact assessments

To evaluate the impact of the ET tanning process on the environment, biochemical analysis of leather processing waste liquor was carried out to assess its ecological risks to the environment [26]. The ET tanning effluent was detected to evaluate the environmental impact parameters of the ET tanning method, and compared with CTDT tanning method and conventional chrome tanning method.It is shown in **Table. 7** thatthe residual Cl^- concentration in wastewater decreased 42.1% in tanning processcompared toconventional chrome tanning method. The emissions of COD and TDS in tanning process reduced by 40.5% and 27.2%, respectively.The COD, TDS and TDC of ET tanning effluent were slightly lower than that of CTDT tanning effluent.Better biodegradability andno neutral salt had been introduced in the process of ET tanningindicated thatthe low environmental administration budget of ET tanningcould benefit the develop sustainably and provide inclined futuristic prospects for cleaner leather manufacturing.

Table.7 Comparison of three tanning methods

Items	Conventional chrome tanning	ET tanning	CTDT tanning
Neutral salt (%)	6	0	0
Chromium (%)	6	0	0
COD_{Cr} (mg/L)	2088.27±16.90	1242±21.20	1443.21±15.7
TDC (mg/L)	15360.13±96.40	8893.60±49.50	9844.40±69.90
TDS (mg/L)	52330.20±211.30	38118±171.12	37554.70±144.80
Undegradable pollutants	Neutral salt and chromium	—	—

Formatting Issue - Best Available Paper/Graphic

4. CONCLUSION

Nowadays, a growing number of countries have implemented strict regulations for promote of sustainable development.Adequacy in achievement of chrome-free tanning is put forward as a new criterion for clean leather manufacturing. In this work, polyether amine triazine derivatives (ET) was prepared via one-step process and a novel chrome-free tanning approach based on ET have beenestablished and evaluatedcomprehensively.ET has been confirmed to combine with the amino group of collagen mainly in the form of covalent bond and produced crosslinking network in the collagen structure, which greatly improved the thermal and physical properties of collagen. Due to the existence of polyether chain in ET structure, part of hydrogen bond and van der Waals force in collagen was broken, loosen fibers, and increased the softness of leather. In addition, the rigidity structure of triazine ring in ET could make a significant contribution to the hygrothermal stability of collagen, after tanning with 8% of ET at pH6 for 6h, the wet-white present shrinkage temperature of 83.9℃ andthickening rateof 87.3%, the environmental impact assessments showed ET tanning shows better environmental protection than conventional chrome tanningas well as CTDT tanning.

Acknowledgments

Thanks also go to Chunxiao Zhang, a researcher in Sichuan University, for technical support.

CRediT authorship contribution statement

Xinju Jia: Conceptualization, Methodology,Formal analysis, Writingoriginal draft.

Ran Tan: Validation, Data curation.

BiyuPeng: Funding acquisition,Supervision

Declaration of competing interest

The authors declare no conflict of interest.

REFERENCES

Formatting Issue - Best Available Paper/Graphic

[1]Bhuiyan MAH, Suruvi NI, Dampare SB, Islam MA et al. Investigation of the Possible Sources of Heavy Metal Contamination in Lagoon and Canal Water in the Tannery Industrial Area in Dhaka, Bangladesh. Environmental Monitoring and Assessment. 2011, 175(1-4):633-649.

[2] Sun X, Jin Y, Lai S, et al. Desirable retanning system for aldehyde-tanned leather to reduce the formaldehyde content and improve the physical-mechanical properties. Journal of cleaner production. 2018, 175(20):199-206.

[3]Li, K. Chen, H. Wang, Y. Shan, Z. Yang, J. Brutto, P. A salt-free pickling regime for hides and skins using oxazolidine. Journal of cleaner production. 2009,17:1603-1606.

[4]Zhang, J. Fan, C. Ai, Z, et al. The synthesis and application of phosphorus-nitrogen flame retardant retanning agent. Leather and Footwear Journal.2015, 15(3):179-188.

[5]Pervaia N. Structural and Performance Characteristics of Pigskin Leather with Different Tanning Methods. Journal of the Society of Leather Technologists and Chemists. 2019, 103(3):122-128.

[6]Zhongbing Lu, X Liao, Shi Bi. The reaction of vegetable tannin-aldehyde-collagen: a further understanding of vegetable tannin-aldehyde combination tannage. Journal of the Society of Leather Technologists and Chemists.2003, 87(5):173-178.

[7]Zengin A, Crudu M, Maier S S, et al. Eco-leather: Chromium-free Leather Production Using Titanium, Oligomeric Melamine-Formaldehyde Resin, and Resorcinol Tanning Agents and the Properties of the Resulting Leathers. Ekoloji. 2012, 21(82):17-25.

[8]Krishnamoorthy, G. Sadulla S, Sehgal P K. Greener approach to leather tanning process: d-Lysine aldehyde as novel tanning agent for chrome-free tanning. Journal of Cleaner Production. 2013, 42:277-286.

[9]Kim, K. H.; Jahan, S. A.; Lee, J. T. Exposure to formaldehydeand its potential human health hazards. J. Environ. Sci. Heal. C 2011, 29 (4):277–299.

[10] Tang R, Wen J, Stote R E, et al. Cyanuric Chloride-Based Reactive Dyes for Use in the Antimicrobial Treatments of Polymeric Materials. ACS Applied Materials & Interfaces, 2020, 13(1):1524–1534.

[11]Klančnik, M. Kinetics of Hydrolysis of Halogeno-S-Triazine Reactive Dyes as a Function of Temperature. Chem. Biochem. Eng. Q. 2008, 22:81-88.

[12]Namazi H, Adeli M. Synthesis of barbell-like triblock copolymers, dendritic triazine-block-poly (ethylene glycol)-block-dendritic triazine and investigation of their solution behaviors. Polymer. 2005, 46(24):10788-10799.

[13]Gholap S, Gunjal N. 2,4,6-Trichloro-1,3,5-triazine (TCT) mediated one pot direct synthesis of N-benzoylthioureas from carboxylic acids. Arabian Journal of Chemistry. 2017, 10(2):1073-1079.

[14]Gotoh Y, Tsukada M, Minoura N. Chemical modification of silk fibroin with cyanuric chloride-activated poly (ethylene glycol): Analyses of reaction site by proton NMR spectroscopy and conformation of the conjugates. Bioconjugate Chemistry. 1993, 4(6):554-559.

[15] Yu L, Qiang X, Cui L, et al. Preparation of a syntan containing active chlorine groups for chrome-free tanned leather. Journal of cleaner production. 2020, 270:122351.

[16] Karimi A R, Tarighatjoo M, Nikravesh G. 1,3,5-Triazine-2,4,6-tribenzaldehyde derivative as a new crosslinking agent for synthesis of pH-thermo dual responsive chitosan hydrogels and their nanocomposites: Swelling properties and drug release behavior. International Journal of Biological Macromolecules. 2017, 105:1088-1095.

Formatting Issue - Best Available Paper/Graphic

[17]AbuchowskiA ,Van Es T , Palczuk N C , et al. Alteration of immunological properties of bovine serum albumin by covalent attachment of polyethylene glycol. The Journal of biological chemistry. 1977, 252(11):3578-3581.

[18] Chen, M.M. Qiang, X.H. Chen, W. Sun, Z. Huang, Q.H. Combination tanningof chrome-free tanning agent F-90 and chrome tanning agent. China Leather.2016, 45:32-36.

[19]Yan Z, Xue W L, Zeng Z X, et al. Kinetics of Cyanuric Chloride Hydrolysis in Aqueous Solution. Industrial & Engineering Chemistry Research. 2008, 47(15):5318-5322.

[20] Yang L, Liu Y, Wu Y, et al. Thermal degradation kinetics of leather fibers treated with fire-retardant melamine resin. Journal of Thermal Analysis& Calorimetry. 2016, 123:413–420.

[21]Krishnamurthy M, Murugan P, Murali A, et al. "Click" Polymer of Carbon Nanotubes for Superhydrophobic Glass and Leather. Green Materials. 2017, 5(2):1-7.

[22] Qiongfen, Xia, Li, et al. Chromium Cross-Linking Based Immobilization of AgNPs Coating on Leather Surface with Broad-Spectrum Antimicrobial Activity and Durability. ACS Applied Materials & Interfaces. 2018,11(2):2352-2363.

[23] Androulaki, K.Chrissopoulou, K.; Prevosto, D. Labardi, M. Anastasiadis, S. H. Dynamics of hyperbranched polymers under confinement: a dielectric relaxation study. ACS Appl. Mater. Interfaces.2015, 7:12387–12398.

[24] A Sionkowska, M Wisniewski, J Skopinska, et al. Molecular interactions in collagen and chitosan blends. Biomaterials. 2004, 25: 795-801.

[25] Song Y, Zeng Y, Xiao K, et al. Effect of molecular weight of acrylic resin retanning agent on properties of leather. The Journal of the American Leather Chemists Association. 2017, 112(4):128-134.

[26] Suresh, V, Kanthimathi, M., Thanikaivelan, P, Rao, J.R, Nair, B.U. An improved product-process for cleaner chrome tanning in leather processing. Journal of cleaner production. 2001, 9, 483-491.

Study on Construction and Properties of Self-cleaning Leather

Dr. Kuang, Wei

Qilu University of Technology
(Shandong Academy of Sciences)

Speaker information-
brief introduction

- **Full-time position: Director of biomass and Tanning Department, Qilu University of Technology, Dr.**
- **Part-time position：**
  - ✓ **Standing member of Technical Committee of China Leather Association.**
  - ✓ **Member of expert bank of China Shoemaking Union.**
  - ✓ **Chief engineer of Jiangsu Dunwang Technology Group Co., LTD.**
- **Technology specialty: functional polymer material, functional leather product.**

Research Background---
Waterproof and antifouling effect picture of traditional leather

Research Background---
Waterproof and antifouling effect picture of traditional leather shoes and boots

Research Background---
Waterproof and antifouling effect picture of self-cleaning leather shoes

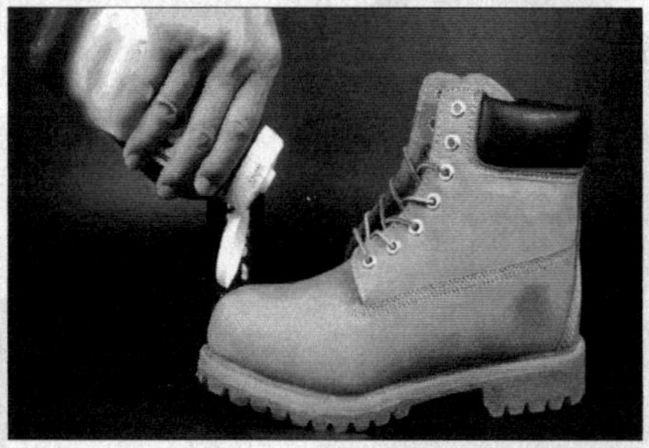

Research Background-Self-cleaning principle

Lotus leaf: "rising unsullied from mud"
——the self-cleaning effect of the surface of lotus leaf

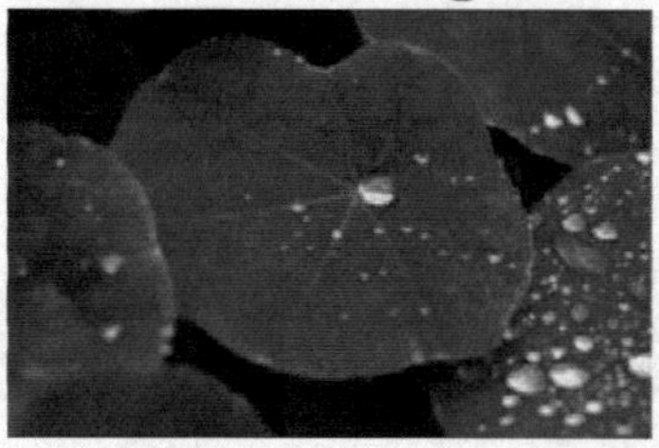

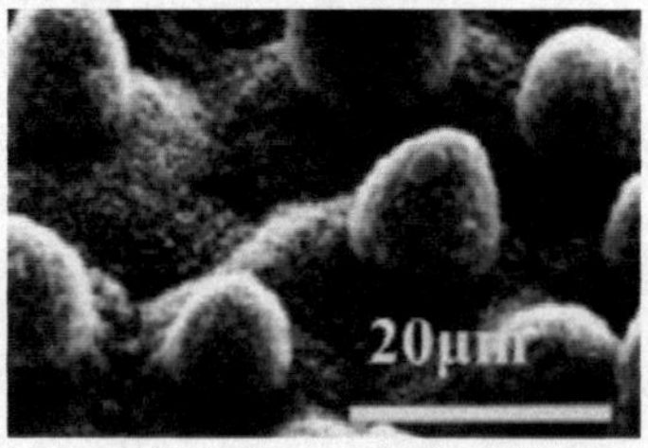

Why?

--- Linear or hairy micron convex bodies and waxy substances with low surface energy on the surface of lotus leaves

Research Background-Self-cleaning plant bionics

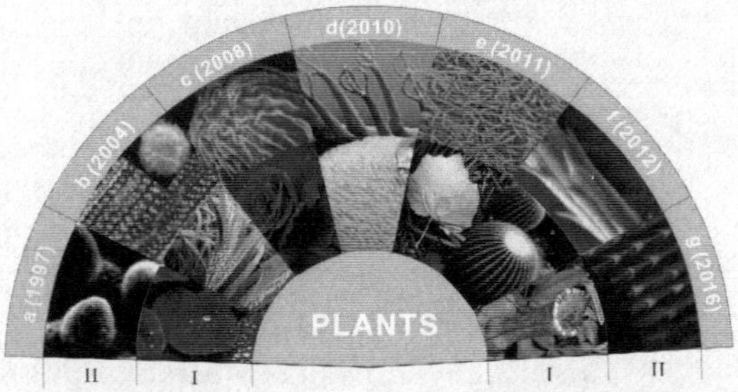

Typical plants (Tape I) with special surfaces and corresponding micro-nano structures (Tape II).
(a) Lotus leaves (b) Rice (c) Red roses (d) Salvinia leaves (e) Poplar leaves (f) Cactus (g) Porphyra

Research Background-Self-cleaning plant bionics

Typical animals(Inner ring) with special surfaces and corresponding micro-nano structures(Outer ring).
(a) Stenocara beetles . (b) Water strider legs . (c) Butterfly wings . (d) Mosquito compound eyes . (e) Fish scale . (f) Spider silk . (g) Gecko feet . (h) Shark skin (i) Clam shell . (j) Earthworm Skin

Construction method of self-cleaning coating - spray coating and wipe coating

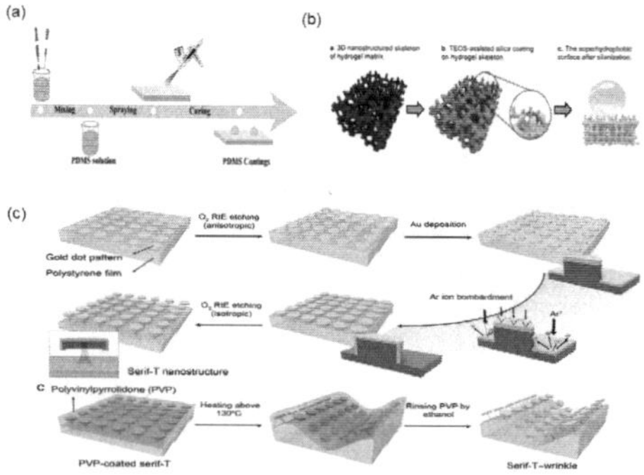

(a) Schematic diagram of the preparation of hydrophobic PDMS coating matrix.
(b)Schematic description of superhydrophobic surfaces templated from a hydrogel matrix.
(c) Preparation steps of a springtail-inspired serif-T shaped textured surface by lithography and wrinkling

Construction method of self-cleaning coating - etching method

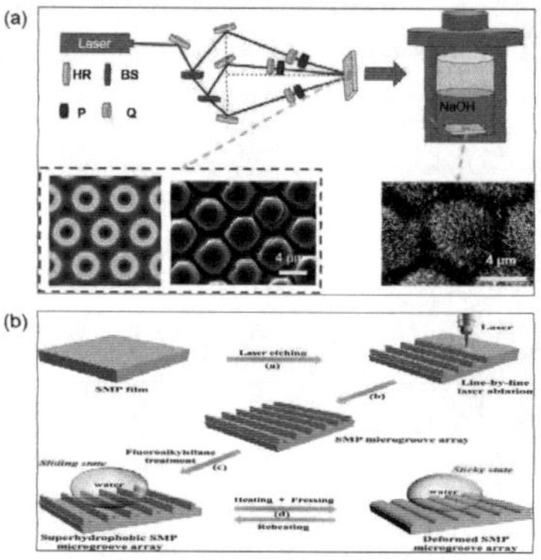

(a) Fabricating bionic taro leaf surfaces by DLIL and hydrothermal treatment.
(b) Schematic diagram of anisotropic superhydrophobic SMP microgroove arrays fabricated by selective femtosecond laser etching.

Completed work and basic experimental conditions

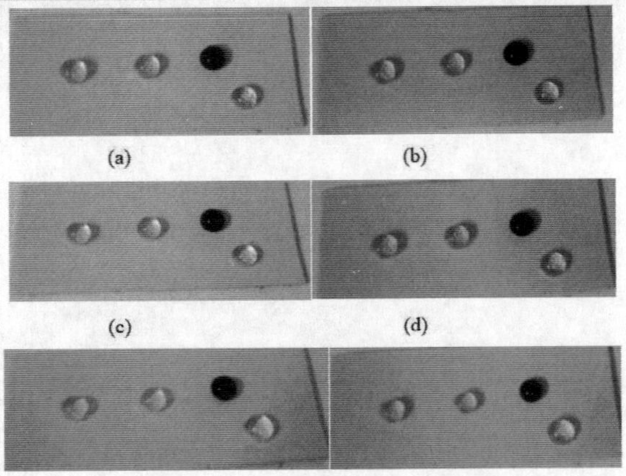

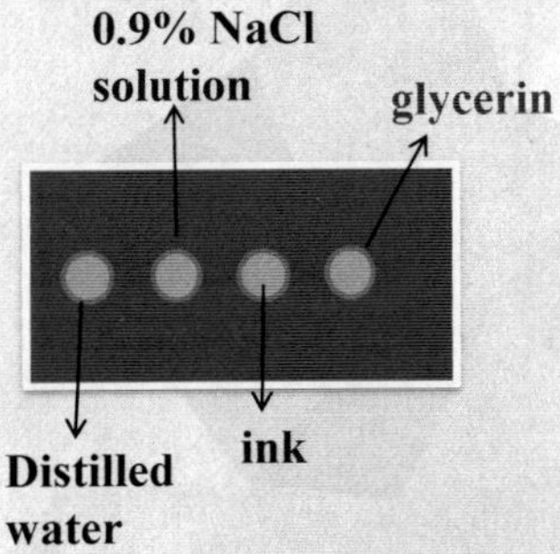

Characterization of coating properties---

determination of static contact angle of self-cleaning vegetable tanned leather

The static contact angle of self-cleaning frosted leather

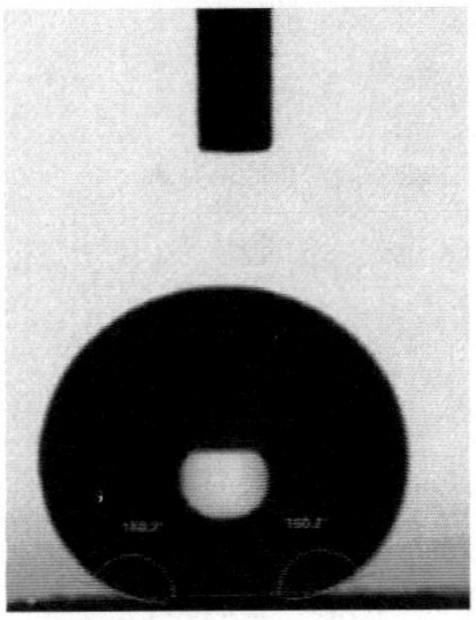

The static contact angle of self-cleaning vegetable tanned leather

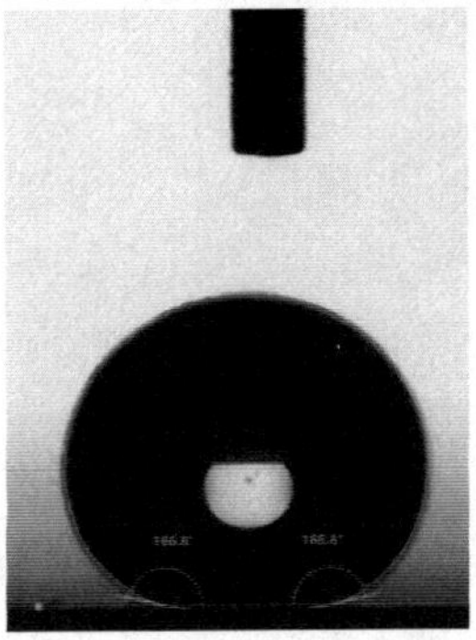

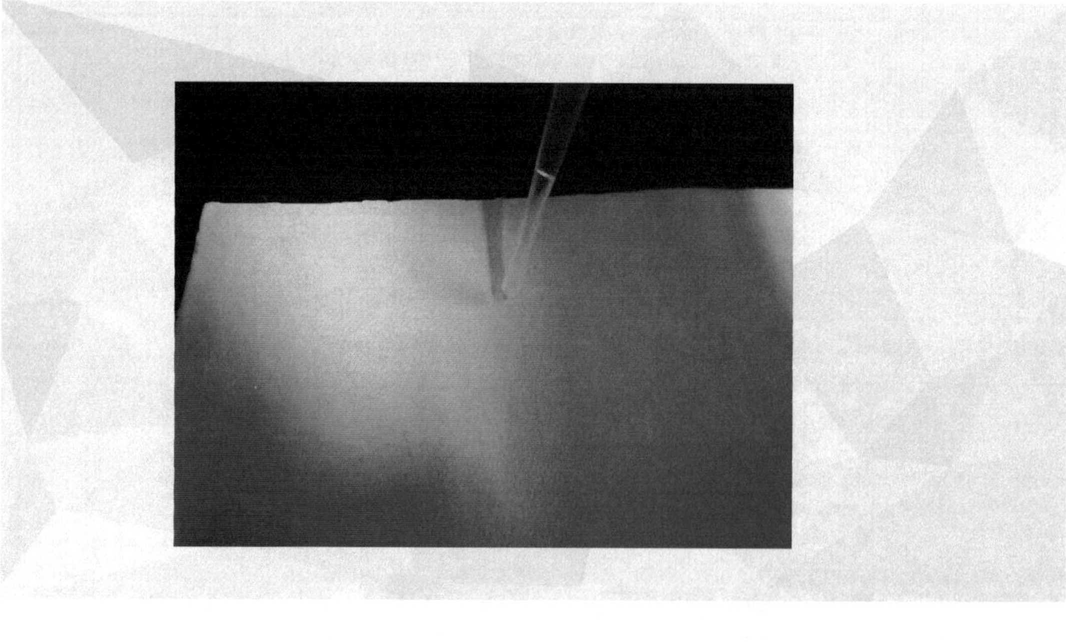

Determination of physical and mechanical properties of leather - Folding fastness

Fold fastness test

1# Uncoated frosted leather
2# Self cleaning wipe coated frosted leather
3# Self cleaning spraying coated frosted leather
4# Uncoated vegetable tanned leather
5# Self cleaning wipe coated vegetable tanned leather
6#: Self cleaning spraying coated vegetable tanned

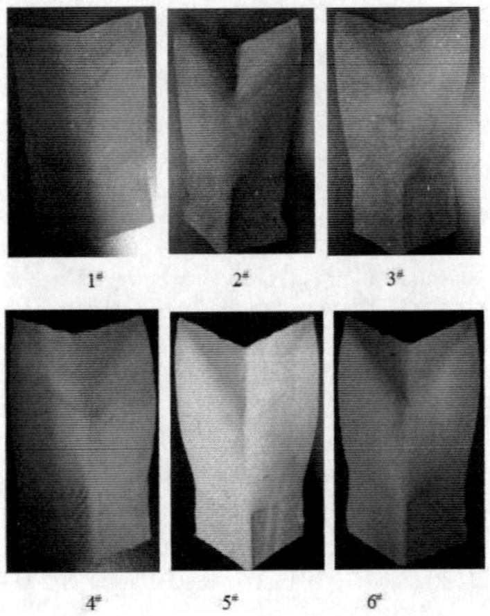

Determination of physical and mechanical properties of leather - moisture permeability

$$P = \frac{76.39 \times m}{d^2 \times t}$$

1#: P=0.3027, 2#: P=0.2936, 3#: P=0.3535
4#: P=0.1927, 5#: P=0.0511, 6#: P=0.0651

- **After self-cleaning coating, the water vapor permeability of leather samples has been influenced.**
- **Like air permeability, the water vapor permeability of frosted leather is generally stronger than that of vegetable tanned leather.**

Prospect:

- ➢ The self-cleaning anti graffiti effect has been realized on the glass substrate.
- ➢ This technology is intended to be used for leather substrate in the future.

Brief introduction of research group

Acknowledgement:

- **Shandong Provincial Key Research and development program**
- **Natural Science Foundation of Shandong Province**
- **Young Doctoral Fund of Qilu University of Technology**

Title

The effect of retanning on the biodegradability of leathers.

Authors

Flowers, K.B.[1] (corresponding author), Dring, E.[1], and Clare, A.[1]

[1]Authenticae Limited, United Kingdom.

Article Objective

The research shows the impact that non-biodegradable chemistry has on the final leather. Leathers from a biodegradable tannage become slower to disintegrate and assimilate when treated with non-biodegradable chemistry compared to biodegradable retannages.

Thematic area

Cleaner technologies for the leather processing

Abstract

A fully biodegradable tannage, using organically masked zeolite, produces a semi-processed tanned leather that fully disintegrates in less than 20 days (when tested using a modified ISO 20200: 2015 disintegration test).

The wet, thinned, biodegradable semi-processed leathers were then treated with chemicals of differing degrees of biodegradability from three retanning chemistry classes (vegetable tannin, zeolite, and syntan).

The results of the research show that as retanning chemical complexity increases, the ultimate biodegradability (as measured by ISO 20136: 2020)decreases.

Considerations of the biodegradability and disintegration, within the results of this research, are linked to the microbiology limitations of current test methods and known data of tanning substances biodegradability.

Introduction

Breakdown of materials will occur differently in the range of biomes that are present on Earth. Substances that are biodegradable in one biome may struggle to achieve the same breakdown in other biomes. Anaerobic environments pose the greatest challenge for organic material biodegradability. Fossil formation, oil formation, and coalification are examples of the difficulty of biodegradability in anoxic sediments that make microbiological growth difficult. Organic substances degrade into fossil carbon sources because they cannot be biodegraded directly into CO_2 and water.

In considering the breakdown of materials it is important to recognise that some substances, that consist of a single chemical type, can be tested for biodegradability and can be generalised into being classed as biodegradable or non-biodegradable (within a specified time frame and biome). However, most modern human materials used for commercial products are composites of substances (leather is a composite of collagen and tanning chemistry). As such a distinction between the substance and the material must be made and the terminology for substances are not easily transferable for use to describe the biodegradability of composites.

There is a difference between the degradation, ultimate biodegradability, and composting of composite materials (Nera, 2020). Ultimate biodegradability is the level of degradation achieved when the test compound is totally utilised by microorganisms resulting in the production of carbon dioxide, water, mineral salts, and new microbial cellular constituents (biomass).

Composting is recognised through an overarching specification in the breakdown of composites and is dependent on the following (BSI, 2006):

1. Biodegradability
2. Disintegration during treatment
3. Effects during treatment

4. Quality of the resulting compost

BS EN 14995, the plastics specification, that is often used to compare the compostability of leather has its limitations (BSI, 2006). The biodegradability test methods recognised by BS EN 14995 do not recognise ISO 20136. ISO 20136 is a leather method and is an aqueous test – the plastic specification recognises solid state respirometry test methods to determine biodegradability (aerobic and anaerobic).

The only leather biodegradability test method is ISO 20136: 2020 which is an aerobic, aqueous biodegradability test that is dependent on tannery or domestic activated sludge as an inoculum. Aerobic aqueous tests rely on bacteria to do the breakdown - which are notoriously incompetent in handling aromatic or other complex chemistry. The test seeks to convert the carbon content of the composite into CO_2 that is measured by respirometry in a bacterial soup.

Carbon present in the sample will be converted to exhaust gases and into carbon locked in as bacterial biomass so it is impossible for the test to reach the end goal of 100% reduction of material into CO_2 and water - hence the end point of the test must be extrapolated. ISO 20136 does not present benchmarks of biodegradability as it is a test method. Specifications like BS EN 14995 go some way to defining what is classified as biodegradable (and what is not) but of course does not include leather (or its chemistry) within its test scope.

The tannage of a leather converts fully biodegradable collagen into a form of material that biodegrades more slowly. Retanning, dyeing, and fatliquoring will add further chemicals which could slow the release of carbon as a nutrient for biodegradation-linked growth. Complex chemistry results in further reductions of carbon recycling by bacteria or fungi. Tanning chemistry used for tanning or retanning can be individually measured as a substance to test its biodegradability (Lofrano*et al.*, 2008).

This work intends to gain some initial insights into how the retanning chemistry affects the aerobic, aqueous biodegradability. The hypothesis to be tested contains the concept that the retanning chemistry does play a role – and insights into the complexity that accompanies that are to be gained.

Materials and Methods

Leather process

The control used in the biodegradability test was the standard hide powder sold by BLC Eurofins, Northampton, UK. The hide powder is made according to the SLTC method SLT 3/7 and is largely Type III collagen.

A whole cattle hide was processed from wet-salted to pickling using a standard beaming process required for zeolite tannage. The hide was divided into quarters, assigned a tannage (and retannage) using a Latin square method. Four leathers would be made from these quarters using the regime outlined in Table 1.

Table 1.

Designation	Retannage	Fatliquor	Dye
Z	None	None	None
1	Zeolite (3%) and protein hydrolysate	Standard	Standard
2	Hydrolysable vegtan (10%) and protein hydrolysate	Standard	Standard
3	Phenolic syntan (3%) and protein hydrolysate	Standard	Standard

All quarters after subdivision and numbering were placed into a standard zeolite tannage. Sample Z had no further retannage performed and was split and shaved (1.8 mm) together with the other three Samples. Sample Z was dried out at room temperature for 3-4 days prior to grinding and then biodegradation testing.

Samples 1-3 were processed using the following shoe-upper regime (see Appendix 1) with the appropriate substitution of the chemistry given above. The leathers were dried and softened as given in Appendix 1. These three samples were then ground prior to biodegradation testing.

Sample preparation for biodegradation and carbon content

Set samples designated for carbon testing were cut into smaller pieces and dried at 105°C till constant weight. These samples and controls were used to measure loss-on-ignition (LOI) by burning 5–10 g of dry material in a muffle furnace at 550°C for 5 h, or longer, until the mass remained constant). The LOI was used as an accurate estimation of the total organic carbon (TOC) content. This form of elemental analysis is common in soils and other materials and can be assumed to be acceptable as a method (this is not well defined in ISO 20136: 2020).

The TOC is used to theoretically calculate the amount of CO_2, $ThCO_2$, that could be evolved by the material in the case of a full biodegradation. ISO 20136 specifies how the $ThCO_2$ is calculated from the following assumptions about the weight percentage of carbon (ISO, 2020).The maximum $ThCO_2$ was calculated for the collagen controls and the test was run to reach a minimum of 70% of CO_2 productionfor those controls to ascertain viability.

Biodegradability testing

The standard method used was the BS EN ISO 20136, leather - determination of the degradability by microorganisms (ISO, 2020).The growth media for the test was formulated as can be seen in Table 2.

Table 2.

Materials	**Quantity (mL)**
Water	788

Inoculum	200
Ferric chloride	2
Magnesium sulfate	2
Calcium chloride	2
Phosphate buffer	4
Ammonium sulfate	2

The CO_2 detection equipment used was an EGA61 respirometer (ADC Bioscientific Limited, Herts, UK) using a multi-channel through-feed infra-red analyser calibrated to 1 ppm resolution. The flow rates were standardised and the time of gas flow through the sensor chamber was optimised.A zero CO_2 channel was used to ensure baseline and the positive control was as described above. The negative control (blank) was the above formulation without a test material.The controls and test samples were run in duplicate.

The blank CO_2 respiration is considered the baseline respiration of the growth medium (See Table 2) and this respiration level was removed from the CO_2 data to reveal the CO_2 being released from the material being tested.

The incubators were shaken on an orbital bed at 150 rotations/minute and connected using leak proof tubing and connectors.

Inoculum

The inoculum used was from the biological tank of the Eurofins BLC tannery in Northampton, UK. The inoculum was obtained on 24h before use, stored in a clean plastic bottle and transported/stored at 4°C. The solids were removed using glass wool.

The biodegradation

The samples were added to the inoculum in the manner outlined by the standard method. Temperatures, CO_2-free oxygen supply, and acclimation of the start of the test were run as per the test method.

Data was recorded on the analyser instrument and the climate control, flow rates, observational effects were checked daily. Any deviations were recorded.

Data preparation and analysis

Raw CO_2 measurements (against the timestamp, recorded on channel number) from the instrument, the flow rates, and the details on the quantities and chemical characteristics of the samples were entered into company software.

The percentage breakdown of the sample as recorded from the CO_2 evolution from the sample was calculated using the method outline in ISO 20136: 2020. The cumulative percentage breakdown is recorded over time and the breakdowns over the 28 days of the test are plotted. The percentage breakdown at the end of the test is recorded as the absolute biodegradation percentage. The absolute biodegradation % is converted into a relative biodegradation % using the following equation:

$$\% \, relative \; biodegradability = \frac{\% \, absolute \; biodegradability \, of \; sample \; x \; 100}{\% \, absolute \; biodegradability \, of \; collagen}$$

Results

The preparatory stages, beaming, tannage, and post tannage proceeded normally as laid out in Appendix 1. The dried leathers were ground into powder for use in the ISO 20136: 2020 test. The carbon content of the samples was determined and are presented in Table 3.

Table 3.

Materials	Carbon (%)	Dry matter (%)
Collagen	50.8	85.0

Tanned	34.63	85.8
Zeolite retan	56.8	85.1
Phenolic syntan	56.6	86.4
Polyphenol	57.0	86.1

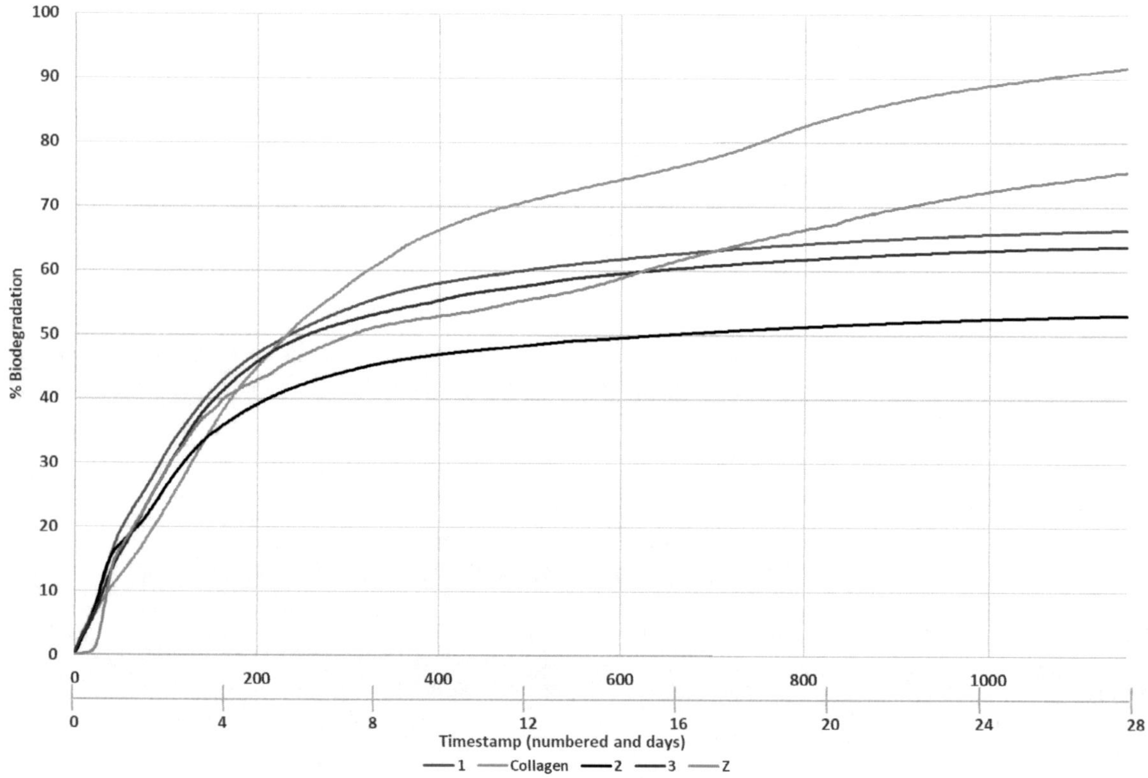
100
90
80
70
60
50
40
30
20
10
0
% Biodegradation
0
200
400
600
800
1000
0
4
8
12
16
20
24
28
Timestamp (numbered and days)
1
Collagen
2
3
Z

Figure 1.

-9-

The biodegradation proceeded as normal, with agitation and aeration (judged by flow) appearing as normal. Material accumulated around the air diffusers and other bacterial biomass aggregatedin the same area. By the end of the biodegradation the solutions appeared lighter coloured. It was difficult to judge a change in material quantity by visual inspection.

Raw data from the respirometer was transferred to company software and the starting masses, carbon quantity, and other chemical characteristics were analysed as per the standard method.The output of the software produced the graph seen in Figure 1 and the calculations from the software provided the data seen in Tables 3 and 4.The mean biodegradation curves show all 5 samples increased as expected with varying rates of CO_2 accumulation per sample.

Discussion

The collagen control biodegradation profile, shown in Figure 1, meant the biodegradation of the test was deemed viable. The collagen profile indicates that the bacteria rapidly breakdown simple proteins, fats, and sugars in the control material. Fluctuations in the CO_2 output indicate the bacteria alter their enzymatic machinery to access different substrates. The cumulation slows down as the secondary metabolism becomes quite costly, carbon that is locked in bacterial biomass also contributes to the slow-down. Collagenase availability (and their metabolic cost) and the chemical complexity of collagen peptides will also present problems for the bacteria.

Zeolite semi-processed material, shown in Figure 1, Sample Z, has a biodegradation profile that is slightly slower than the collagen control. The tannage that is typical in zeolite tannage does not appear to pose a difficulty to collagenolytic bacteria. Lack of CO_2 production is linked to decreased biological assimilation There appears to be increasing assimilation difficulties as the retanning type moves from collagen, through semi-processed material (no retannage), and on to the three other retannage types.

The semi-processed material curve lags behind the other materials that received a retannage in the initial 15 days of biodegradation. After 15 days the CO_2 release in Sample Z increased. The increase is most likely due to a bacterial population that has matured into a critical mass that can start breaking down the nutrients available for assimilation, or the breakdown of the materials have reached the point where assimilation becomes easier.

Figure 1also shows mean curves of the cumulative percentage biodegradation shown for all 3 samples that received a retannage. The curves show that as the chemistry complexity in the material moves from semi-processed (tanned) into an increased quantity of zeolite retannage (Sample 1), through a phenolic syntan retannage (Sample 3), into a hydrolysable polyphenol retannage (Sample 2) – the biodegradation rate slows down.

Table 4.

Samples	Absolute biodegradation (%)	Relative biodegradation (%)
Collagen control	91.59	100.00
Tanned	74.42	81.25
Zeolite retan	66.36	72.42
Phenolic syntan	63.82	69.68
Polyphenol	53.14	58.02

Table 4 supports the view outlined by the biodegradation curves seen in Figure 1. The absolute and relative biodegradation percentage breakdown supports the pattern that increased aromaticity (and aromatic complexity) does cause a lower biodegradation profile.

It must be reiterated that these biodegradation profiles are for aerobic, aqueous biodegradation. Aqueous environments lack fungal species that are better at handling aromatic chemical species. Biodegradation, specifically disintegration associated with breakdown, are not easily generalised to any biome. Partial breakdown into disintegrated products (but not assimilated) may produce less CO_2 and may better build the organic matter of soils.

Sample 2, which was a hydrolysable polyphenol showed decreased assimilation and CO_2 release. The anti-bacterial properties of polyphenols have recently been highlighted (Poles *et al.*, 2019). This may be a contributor to the decreased biodegradability and will be the focus of future studies. Polyphenols are not easily degraded by bacteria, which is often a cause for their high chemical oxygen demand (Elmansour*et al.*, 2020). Celluloses, hemi-celluloses, lignins (in particular) and complex aromatics often require fungal enzymatic systems or abiotic factors (like aeration/oxidation) that will chemically degrade the ring prior to further hydrolysis and assimilation.

The results shown in Table 4 and Figure 1 do suggest that the syntans are less biodegradable than zeolite retannage but are more degradable than polyphenols. Future studies will focus on petroleum-based polymers that could present difficulties to aqueous biodegradability. This study doesn't demonstrate that petroleum-based phenolic syntans (that are comprised of fossil carbon) compared to polyphenolic biogenic carbon are more environmentally friendly as the CO_2 respirometry doesn't differentiate between the two forms of carbon. This must be the focus of future studies.

Conclusion

The initial results of the research show that as retanning chemical complexity (aromaticity and chemical/collagen interaction) increases, the ultimate biodegradability (as measured by ISO 20136: 2020)decreases.

The conclusion that can be drawn from this that in an aquatic biome (dominated by bacteria), the leather becomes harder to breakdown as the retannage becomes more complex. Retanning products with a higher COD may be an early indicator that as a constituent it could pose biodegradability difficulties (future studies will confirm this).

Acknowledgements

The authors would like to thank the organisers and sponsors of the 2020 IULTCS conference. The authors declare no conflicts of interest and that this research was self-funded.

References

Nera (2020) Biodegradability and disintegration of leather. A Whitepaper from Nera Tanning, Weesp, The Netherlands. p.1-11.

BSI. 2006. BS EN ISO 14995:2006 Plastics - evaluation of compostability - test scheme and specifications. Milton Keynes: British Standards Institute.

ISO, 2020. Leather — determination of degradability by micro-organisms(ISO 20136: 2020). International Standards Organisation, Geneva, Switzerland.

Lofrano, G., Aydin, E., Guida, M. and Belgiorno, V. (2008) Characterization, fluxes, and toxicity of leather tanning bath chemicals in a large tanning district area (IT).*Air, & Soil Pollution: Focus***8**(5):529-542.

Poles, E., Polissi, A., Battaglia, A., Giovando, S., and Gotti, M. (2009) Study on the antibacterial properties of leathers tanned with natural tannins and their interactions with shoes inhabiting bacteria. *Proceedings of the XXXVth IULTCS Congress, Dresden.* p. 1-10.

Elmansour, T.E, Mandi, L., Ahmali, A., Elghadraoui, A., Aziz, F.,Hejjaj, A., Del Bubba, M., and Oazzani, N. (2020) Effect of polyphenols on activated sludge biomass during the treatment of highly diluted olive mill wastewaters: biomass dynamics and purifying performances. *Water & Science Technology***82**(7): 1416-1429.

Appendix 1

Wet-back

200%	Water@35°C		
0.5%	Formic acid (1:10)	20'	pH=
Drain			

Neutralise

100%	Water@30°C		
1%	Sodium formate	15'	
+0.5%	Sodium bicarbonate (1:3)		in 2 x 0.25% additions
1.5%	Neutralising syntan	45'	pH =
Drain, wash, drain			

Retan

70%	Water@35°C		
X%	Retanning chemical given in Table 1		
6%	Biopolymer (protein)	30'	
+3%	Dye (1:20)		
5%	Sulfited natural fatliquor	30'	Check cut

Fix

+1.5%	Formic acid (1:10)	30'	in 3 x 0.5% additions
Drain			

Contd...

Appendix 1 contd...

Fatliquor

100%	Water@45°C			
0.3%	Sodium bicarbonate (1:3)		10'	
+3.5%	Sulfited fish oil			
3.5%	Lecithin fatliquor			
1%	Sulfited neat's foot oil (Blend, dilute 1:5)			
6%	Biopolymer (protein)		30'	
+3%	Dye (1:20)			
5%	Sulfited natural fatliquor		30'	Check cut

Fix

+1.2%	Formic acid (1:10)	20'	in 3 x 0.4% additions

Drain

Wash

100%	Water@35°C	
2%	Melamine	10'

Drain, rinse, horse up, set-out, vacuum dry, condition, vibration stake

Formatting Issue - Best Available Paper/Graphic

Microbial degradation of animal hair and preparation of organic compost thereof

NumbiRamuduKamini[a], Parthasarathy BaskaranSujiritha[a], ShakilaShobanaTheagarajan[b] and Palanivel Saravanan[c]

[a]Department of Biochemistry and Biotechnology, CSIR-Central Leather Research Institute, Chennai, 600020, Tamil Nadu, India

[b] Project Planning & Business Development (PPBD), CSIR-Central Leather Research Institute, Chennai, 600020, Tamil Nadu, India

[c] Cell for International Consultancy & Social Impact Assessment, CSIR-Central Leather Research Institute, Chennai, 600020, Tamil Nadu, India

* Corresponding author.

Phone: +914424437212

Fax: + 91 44 24911589

E-mail address: nrkamini@clri.res.in

Abstract

About 50 tons of hair waste is generated from the tanning industries on processing of every 1000 tons of hides/skins. The disposal of hair wastes results in many significant negative environmental impacts.Keratin has wide range of application in agricultural, cosmetic and pharmaceutical industries. The keratin rich hair waste is recalcitrant and its degradation is laggard. CSIR-CLRI developed a bioprocess technology for efficient degradation of hair using a novel bacterial strain, *Brevibacteriumluteolum* MTCC 5982 in liquid medium containing salts and animal hair. The process was standardized in flask level and scaled up to 500 L in 750 L fermentor. The resulting fermentation liquor was used for the manufacturing of organic compost. The process could be completed within a time period of 10 to 12 days as against the conventional composting period of 90 days. Pilot scale production of compost has been standardized and the efficacy of the product has been tested at field level. The compost could serve as an effective organic fertilizer and could increase the yield of paddy by 20±5% and horticultural crops of bhendi, chilli and amaranthus by 25±5%. The process provides viable technological solutions to address the issue of tannery hair waste disposal on one hand and ensure financial returns on the other hand. The bulk production of compost from hair wastes could reduce the import and consumption of chemical fertilizers. This is an exemplary model of circular economy as the waste of the leather industry serves as an organic fertilizer of the agricultural industry.

Keywords: Hair degradation, organic compost, keratinase, keratin hydrolysate, agricultural industry, tannery.

Formatting Issue - Best Available Paper/Graphic

1. Introduction

Leather industry utilizes raw hides/skins, by-product of meat industry, as the raw material.About 6.6 million tons of raw hides from bovine animals and 0.8 million tons of skins from ovine animals are converted into leather world over every year. Hair wastes of about 56 kilotons is generated annually from the tanneries producing ovine skins (Food and Agriculture Organization of the United Nations, 2016; Thankaswamy et al., 2018). Poor management of hair wastes by dumping, burial and incinerationnot only pollute the soil, air and underground water resources, but also may cause different human infirmities (Jana et al., 2020). Keratins are recalcitrant, stable, water-insoluble structural fibrous proteins and its degradation is slow in nature. The poor degradability of keratin can be attributed to the stabilization of the cuticle of hair due to the high degree of cross linkages by disulphide and hydrogen bonds.

Conventional methods of hair degradation include the physico-chemical methods, which are energy intensive, economically not feasible and also results in the destruction of essential amino acids (Nnolim and Nwodo, 2020). Microbial and biotechnological approach could be a promising alternative for disposing or utilizing keratinous wastes using the enzyme, keratinase. Keratinases are non-specific serine type proteases that have many applications in several sectors viz., leather, textiles, detergents, cosmetics, prion degradations, in production of biodegradable films, glues and foils and agro-industrial waste degradation (Barman et al., 2017; Dada and Wakil, 2021). There have been many reports on degradation of animal hair and feather using various microorganisms, including bacteria such as *Bacillus licheniformis, Brevibacillus brevis, Stenotrophomonas maltophilia*andyeast strain, *Trichosporonloubieri* (Thankaswamy et al., 2018). However, the microorganisms required addition of nitrogen sources, such as yeast extract, peptone, soybean meal, ammonium ions and soy flour for the growth and production of keratinase (Gafar et al., 2020)and the time required for degradation of hair was significantly high. Therefore, efficient degradation of hair in the shortest duration without addition of nitrogen source is essential to utilize the keratin towards production of valuable keratin hydrolysate for various applications.

Due to the growing demand for organic food, the requirement for bio-fertilizers has been increasing significantly. Previous reports on the preparation of organic fertiliser from keratinous waste is associated with alkaline treatment (Gousterova et al., 2003; Nustorova et al., 2006) or pre-treatment of the substrate by the methods such as chopping, cutting and grinding (Falco et al., 2019) and the time required for the preparation of the product was about 90 days.

The aim of the present study is to degrade the animal hair waste efficiently without any pre-treatment of the raw material using a novel bacterial strain, *Brevibacteriumluteolum*MTCC 5982 and to prepare organic compost with the resulting keratin hydrolysate in a short duration of time. The preparation of organic compost was made by mixing appropriate volumes of the keratin hydrolysate with agro industrial residues and soil. The process was scaled up at pilot level. Throughout the composting process, the physico-chemical properties of the composting material were assessed at regular intervals. The effect of organic compost on growth and yield of various agricultural crops such as paddy (*Oryza sativa*), amaranthus (*Amaranthus viridis*), chilly (*Capsicum annuum*)and bhendi (*Abelmoschus esculentus*) was evaluated in field trials at different districts of Tamilnadu, India.

2. Materials and methods

2.1. Materials

Hair waste was obtained from Dindigul leather cluster, Tamilnadu, India, where wet-salted goatskins are processed into tanned leathers. Hair waste was cleaned by washing with water and air dried. All the chemicals used in the present study were of AR grade and purchased from Hi-Media Limited and S.D. Fine Chemicals Limited, Mumbai, India. Wheat bran and saw dust were purchased from local market, Chennai, India. Soil and dried leaves were collected from the CSIR- CLRI grounds, Chennai, India.

2.2. Degradation of animal hair

The keratinolytic organism, *Brevibacteriumluteolum* MTCC 5982 isolated from the hair dump yard of Dindigul Tannery, Tamil Nadu was used for the degradation of hair(Thankaswamy et al., 2018). The strain was maintained on nutrient agar slants at 4 ˚C. The organism was grown on nutrient broth for 24 h at 30 ˚C and used as preinoculum. Hair degradation was initially carried out in 250 ml Erlenmeyer flasks containing 50 ml of basal salt medium containing 0.3 g/L of K_2HPO_4, 0.4 g/L of KH_2PO_4, 0.5 g/L of NaCl and 40 g/L of goat hair using 1.0 % (v/v) preinoculum for 72 h at 30 ˚C. The process was scaled up in 75 L fermentor (Scigenics, India) with working volume of 50 L containing the basal salt medium and 4 % of goat hair. The medium was sterilized *in-situ* at 121 ˚C for 20 min and inoculated with 10 % (v/v) of preinoculum grown in nutrient broth for 24 h. The agitation speed and aeration were set at 200 rpm and 2.0 vvm, respectively and fermentation was carried out at 30 ˚C for 96 h. Samples were withdrawn at 6 h intervals and analysed for the keratinase activity and degree of hydrolysis. The parameters like pH and dissolved oxygen in the fermentation broth were monitored throughout the process. Further, scale-up of hair degradation was carried out in 750 L fermentor (Scigenics, India) with working volume of 500 L and all other conditions remained the same. The fermentation broth (keratin hydrolysate) was collected and used for compost preparation.

2.3. Enzyme assay and degree of hydrolysis

The samples collected from the fermentation culture mediumwere centrifuged at 8000 rpm for 10 min at 4 °C and the keratinase activity was determined using keratin azure as substrate (Thankaswamy et al., 2018). One unit (U) of keratinase activity was defined as the amount of enzyme required to bring about an increase in absorbance of 0.01 at 595 nm (Cai et al., 2008). The degree of hydrolysis (DH) was determined using TCA (Hoyle and Meritt, 1994). To 1 mL of culture supernatant, 1 mL of 20% TCA was added and centrifuged at 10000 rpm for 10 min at 4 °C to collect 10% TCA soluble fraction. Total nitrogen present in 10% TCA soluble fraction was estimated by Elemental CHNS analyser. The degree of hydrolysis was calculated using the formula,

$$\text{DH}\,\% = \frac{\text{Total soluble nitrogen content in 10\% TCA X 100}}{\text{Total nitrogen content}}$$

2.4. Preparation of compost

Compost was prepared in triplicates at lab-scale by mixing the fermentation liquor (5 L of hair hydrolysate) and optimized ratio of 5 kg of composting material (soil – 2.0 kg; pulverized dry leaves – 1.5 kg; saw dust – 1.0 kg and wheat bran – 0.5 kg). The mixture was transferred to the trays and kept at ambient temperature (30 ± 2 °C) for 10-12 days. The moisture content was maintained at 55 ± 2% throughout the process by sprinkling water. Samples were collected from the trays at 24 h intervals to determine the pH, temperature, moisture content, total nitrogen, organic carbon and CFU/g. The moisture content was determined with a moisture analyser based on weight loss at 120 °C, ending with a constant weight (Moisture analyzer, Radwag MA 50.R). The pH measurements were carried out with aqueous suspensions of the compost samples (1:6, w/v, compost/water ratio) using a pH electrode (Cyberscan pH Tutor Meter). Quantitative estimation of culturable aerobic bacteria was determined by inoculating 0.1 ml of different tenfold serial dilutions of compost samples on the nutrient agar plates and incubated at 30 °C for 72 h. The number of organisms were expressed as colony forming units (CFU) per gram dry weight of compost. After 12 days of composting, the compost was analysed for its pH, electrical conductivity (mS/cm), total nitrogen (%), phosphate P_2O_5 (%), potassium as K_2O (%), organic carbon (%) and C/N ratio at M/s. National Agro Foundation (NAF) Ltd., Tharamani, Chennai. The prepared compost was evaluated at lab scale (10-15 %) on the growth and yield of chilly plant.

The compost preparation was scaled up to 500 kg level. The keratin hydrolysate of 500 L was mixed with 500 kg of composting material comprising of soil (200 kg), pulverized dry leaves (150 kg), saw dust (100 kg) and wheat bran (50 kg) in 10 batches (50 kg/batch) using ribbon blender mixer (Thirumalai Industries, India). The resulting mixture was transferred to composting pits with following dimensions of 5 X 2.5 X 1.5 m (length X width X height) (Fig. 1) and mixed well at 12 h interval from top to bottom. The moisture content was maintained at 55 ± 2 % by sprinkling water. Temperature of the composting material was measured regularly using a glass thermometer. Samples were collected at 24 h interval and analysed for elemental analysis, pH and CFU/g. The process was continued until a target C/N ratio of less than 20 was attained. Thereafter, the compost was collected and dried. The characteristics of compost were evaluated by NAF for field application studies.

Fig. 1. Preparation of organic compost at pilot level (500 kg)

2.5. Field application studies

Field application studies were conducted by NAF at Illedu village, Kanchipuram District for paddy crop, bhendi and amaranthus and at Ramanathapuram District for chilly crop using the prepared compost. Randomized block design (RBD) was used for the application of compost ranging from 250 kg to 1000 kg / acre. The compost was sprinkled in the field before transplantation of seedlings. The growth and yield of the crop was assessed and compared with the yield of crops that were grown without compost.

3. Results and discussion

3.1. Degradation of animal hair

The *Brevibacteriumluteolum* MTCC 5982 degraded the animal hair in shake flasks at 72 h with production of 98.85 U/mL of keratinase and 80% degree of hair hydrolysis

(Thankaswamy et al., 2018). The keratinase produced by the bacterium hydrolysed the complex supramolecular organization of keratin. The degradation of hair is primarily due to the cleavageof –S–S– bond in cystine present in keratin to form cysteine by the keratinase enzyme (Fig. 2) (Paul et al., 2016). The disulfide bond in cystine is primarily responsible for the structural integrity of hair.

Fig. 2. Structural and molecular breakdown of keratin by keratinase

The process of hair degradation was scaled up in 75 L fermentor using the optimized production medium. A maximum keratinase activity of 100 ± 2 U/mL and 81.5 % degree of hydrolysis was obtained at 66 h of fermentation (Fig. 3a). The hair degradation was further scaled up to 750 L fermentor with a working volume of 500 L under the optimum conditions. The keratinase productivity was 1.50 and 1.65 U/ml/h (at 66 h and 60 h of fermentation) using 75 L and 750 L fermentors, respectively. The degree of hydrolysis increased constantly with increase in keratinase production and a maximum of 82% hydrolysis was attained in 750 L fermentor. The profile of dissolved oxygen and pH of the fermentor results are shown in Fig.

3b.During fermentation, the pH increased gradually from 7.0 to 8.4 and 8.9 in 750 and 75 L fermentor, respectively. The tendency towards alkalization of the medium was due to the hydrolysis of hair and accumulation of NH_4^+ ions and the change in pH was an indicator for degradation of keratin (Maruthi et al., 2011).

Fig. 3. Scale up of hair degradation at 50 L and 500 L level in fermentor. (a) Degree of hydrolysis and keratinase production and (b) Profile of pH and dissolved oxygen during degradation.

Complete solubilization of hair within72 h indicated that the organism utilizes hair as the sole source of carbon and nitrogen with maximum production of keratinase (Williams et al., 1990). The degree of hydrolysis was 82% and the presence of amino acids in the keratin hydrolysate indicated that the integrity of keratin was completely lost in 72 h (Thankaswamy et al., 2018). Thus, the keratin hydrolysate (fermentation broth) could be used as nitrogen source for the preparation of compost.

3.2. Preparation of compost

3.2.1. Physico-chemical properties and microbiological analysis

The changes in pH, temperature, moisture content, total nitrogen, organic carbon and bacterial count during composting process are shown in Table 1. Temperature measurements throughout the composting process provided an adequate real-time indication ofmaintenance of ideal conditions that support microbial degradation (Said-Pullicino et al., 2007). The temperature increased rapidly from 33 to 45±2°C on second day of the process and turning of the composting materials at 12 h intervals enhanced the aeration and decreased the temperature. The increase in temperature could be due to the utilization of organic matter by the bacterial population as reported by Liu et al. (2011). The temperature gradually increased to 48±2°C on the 8th day and thereafter decreased to 34±2 °C at the end of composting process. The initial pH of the composting matrix was pH 7.6±0.2 and gradually increased to pH of 8.2 as the degradation progressed and remained at 7.4±0.2 on the 14th day. The moisture content was adjusted to 55 ± 2% throughout the process, since microbial activity is limited at lower moisture levels. Moisture loss during the composting process could be considered as an index of decomposition rate because the heat generation leads to vaporization (Liao et al., 1997). Similar trend was observed at pilot level production of compost. The population of total culturable aerobic bacteria was initially $18 \pm 4 \times 10^3$ CFU/g and increased to $205 \pm 10 \times 10^3$ CFU/g at the end of composting.

3.2.2. Characteristics of compost

The compost prepared at 5, 50 and 500 kg levels were analysed at National Agro Foundation (NAF)for elemental analysis and compared with the acceptable limits of city compost. The results are shown in Table 2. Initially, the organic carbon was found to be 45.12 % and 46.52 % in 5 kg and 500 kg level of compost, respectively. It decreased to 23.18 % and 24.80% as the degradation progressed. During the composting process, carbon dioxide was mainly emitted because of the degradation of complex organic molecules (Guo et al., 2012). The total nitrogen content was 1.22% and 1.29 % in 5 kg and 500 kg level of compost, respectively at the end of composting process. The loss of nitrogen through the formation of ammonia resulted in decrease of total nitrogen contentduring the initial 4 days of composting. The initial C/N ratio of 38.23 and39.42was decreased to 19.23and 19.75at the end of composting (12th day) at 5 and 500 kg level, respectively. The C/N ratio represents the forms of organic carbon and nitrogen in the composting material that are more susceptible to biological degradation and the decrease in ratio indicated the stability of composting process (Said-Pullicino et al., 2007).

3.3. Field application studies

The results on application of compost at laboratory scale significantly enhanced the growth and yield of chilly crops. Further, field studies were conducted to evaluate the influence of compost on growth and yield of paddy, amaranthus, chilly and bhendicrops and the results are shown in Table 3. The RBD used for the application of compost indicated that use of 500 kg/acre of compost along with required dose of fertilizer significantly increased the growth and yield of the crops compared to the control fields, which were without the compost. The increase in yield of various crops were 20.0, 34.5, 31.0 and 19.5 % with paddy, amaranthus, bhendi and chilly crops, respectively (Table 3). This increase could be due to the addition of fermentation liquor containing the bacterium and the protein rich keratin hydrolysate in the composting materials, which may accelerate the degradation of organic matter along with the soil microflora in attaining the ideal C/N ratio of the product. Moreover, the addition of compost to the soil could increase the water holding capacity and other mineral content of the soil as reported by Tamreihao et al. (2019) using feather hydrolysate. The enhanced yield of crops would lead to the increase in profit for the farmers thereby leading to the improvement in their livelihood.

4. Conclusion

In the present study, the degradation of animal hair waste using keratinase producing *Brevibacteriumluteolum* MTCC 5982 at pilot scale in 500 L fermentor was standardized. The fermentation liquor containing the bacterium and the keratin hydrolysate is utilized for compost preparation without discharging any waste to the environment. The addition of compost enhances the yield of crops and reduces the consumption of the chemical fertilizer to about one-third. The process provides a sustainable solution for the disposal of hair waste of the tanning industries and benefit the agricultural industry by reducing the import and ensure financial returns, thereby stands alone as an exemplary model for circular economy.

Acknowledgements

The authors thank Director, CSIR-CLRI, Chennai, India for his kind permission to present this work and CSIR, India for financial support to carry out the work under the 12th Five-year plan project ZERIS-CSC0103-WP06.

Formatting Issue - Best Available Paper/Graphic

References

Barman, N.C., Zohora, F.T., Das, K.C., Mowla, M., Banu, N.A., Salimullah, M. and Hashem, A., (2017): *Production, partial optimization and characterization of keratinase enzyme by Arthrobacter sp. NFH5 isolated from soil samples*. AMB Express., 7,1-8

Cai, C., Chen, J., Qi, J., Yin, Y. and Zheng, X., (2008): *Purification and characterization of keratinase from a new Bacillus subtilis strain*. Journal of Zhejiang University SCIENCE B., 9,713–720

Dada, M.T. and Wakil, S., (2021): *Hair removal potential of keratinase produced by Bacillus species isolated from feather waste*. Journal of Microbiology, Biotechnology and Food Sciences., 11, e2113

Falco, F.C., Espersen, R., Svensson, B., Gernaey, K.V. and Lantz, A.E., (2019): *An integrated strategy for the effective production of bristle protein hydrolysate by the keratinolytic filamentous bacterium AmycolatopsiskeratiniphilaD2*. Waste Management., 89, 94-102

Food and Agriculture Organization of the United Nations (FAO), (2016): World statistical compendium for raw hides and skins, leather and leather footwear, 1999-2015

Gafar, A., Khayat, M.E., Ahmad, S.A., Yasid, N.A. and Shukor, M.Y., (2020): *Response surface methodology for the optimization of keratinase production in culture medium containing feathers by Bacillus sp. UPM-AAG1*. Catalysts., 10, 848

Gousterova, A., Nustorova, M., Goshev, I., Christov, P., Braikova, D., Tishinov, K., Haertle, T. and Nedkov, P., (2003): *Alkaline hydrolysate of waste sheep wool aimed as fertilizer*. Biotechnology & Biotechnological Equipment., 17, 140-145

Guo, R., Li, G., Jiang, T., Schuchardt, F., Chen, T., Zhao, Y. and Shen, Y., (2012): *Effect of aeration rate, C/N ratio and moisture content on the stability and maturity of compost.* Bioresource Technology., 112, 171-178

Hoyle, N.T. and Merritt, J.H., (1994): *Quality of fish protein hydrolysates from herring (Clupeaharengus).* Journal of Food Science., 59, 76–79

Jana, A., Halder, S.K., Dasgupta, D., Hazra, S., Mondal, P., Bhaskar, T. and Ghosh, D., (2020): *Keratinase Biosynthesis from Waste Poultry Feathers for Proteinaceous Stain Removal.* ACS Sustainable Chemistry and Engineering., 8, 17651-17663

Liao, P.H., Jones, L., Lau, A.K., Walkemeyer, S., Egan, B. and Holbek, N., (1997): *Composting of fish wastes in a full-scale invessel system.* Bioresource Technology., 59, 163-168

Liu, D., Zhang, R., Wu, H., Xu, D., Tang, Z., Yu, G., Xu, Z. and Shen, Q., (2011): *Changes in biochemical and microbiological parameters during the period of rapid composting of dairy manure with rice chaff.* Bioresource Technology., 102, 9040-9049

Maruthi, Y.A., Lakshmi, K.A., Rao, S.R., Chaitanya, D.A., (2011): *Degradation of feather and hair by Chrysosporiumtropicum: a potent keratinophilic fungus.* African Journal of Biotechnology., 10, 3579–3584

Nnolim, N.E. and Nwodo, U.U., (2020): *Bacillus sp. CSK2 produced thermostable alkaline keratinase using agro-wastes: keratinolytic enzyme characterization.* BMC Biotechnology., 20, 1-14

Formatting Issue - Best Available Paper/Graphic

Nustorova, M., Braikova, D., Gousterova, A., Vasileva-Tonkova, E. and Nedkov, P., (2006): *Chemical, microbiological and plant analysis of soil fertilized with alkaline hydrolysate of sheep's wool waste*. World Journal of Microbiology and Biotechnology., 22, 383-390

Paul, T., Jana, A., Mandal, A.K., Mandal, A., Mohpatra, P.K.D. and Mondal, K.C., (2016): *Bacterial keratinolytic protease, imminent starter for NextGen leather and detergent industries.* Sustainable Chemistry and Pharmacy., 3, 8-22

Said-Pullicino, D., Erriquens, F.G. and Gigliotti, G., (2007): *Changes in the chemical characteristics of water-extractable organic matter during composting and their influence on compost stability and maturity*. Bioresource Technology., 98, 1822-1831

Tamreihao, K., Mukherjee, S., Khunjamayum, R., Devi, L.J., Asem, R.S. and Ningthoujam, D.S., (2019): *Feather degradation by keratinolytic bacteria and biofertilizing potential for sustainable agricultural production*. Journal of Basic Microbiology., 59, 4-13

Thankaswamy, S.R., Sundaramoorthy, S., Palanivel, S. and Ramudu, K.N., (2018): *Improved microbial degradation of animal hair waste from leather industry using Brevibacteriumluteolum (MTCC 5982).* Journal of Cleaner Production., 189, 701-708

Williams, C.M., Richter, C.S., MacKenzie Jr, J.M. and Shih, J.C., (1990): *Isolation, identification and characterization of a feather-degrading bacterium.* Applied and Environmental Microbiology., 56, 1509-1515

Formatting Issue - Best Available Paper/Graphic

Table 1

Profile of different parameters during composting process at pilot level (500 kg)

Parameters	Days							
	Initial	2	4	6	8	10	12	14
pH	7.6±0.2	7.9±0.2	8.0±0.2	8.2±0.2	7.8±0.2	7.6±0.2	7.6±0.2	7.4±0.2
Temperature (°C)	33±2	45±2	47±2	47±2	48±2	41±2	35±2	34±2
Moisture (%, w/w)	55±2	56±2	55±2	55±2	54±2	54±2	55±2	55±2
CFU/g($\times 10^5$)	18±4	47±7	115±10	162±8	190±10	198±12	209 ±10	205±10
Organic carbon (%)	46.52	41.91	35.55	31.60	31.09	28.08	25.49	24.80
Total nitrogen (%)	1.18	1.10	1.07	1.19	1.22	1.25	1.29	1.29
C/N ratio	39.42	38.10	33.22	26.55	25.49	22.46	19.75	19.22

Formatting Issue - Best Available Paper/Graphic

Table 2:

Analysis of compost prepared at 5, 50 and 500 kg level

Parameters	Fertilizer control (Acceptable limit for city compost)	5 kg	50 kg	500 kg
pH	6.5-7.5	7.17	7.46	7.3
Conductivity (mS/cm)	Less than 4.0	1.46	1.51	1.67
Moisture (%)	15.0-25.0	15.80	17.36	16.21
Total nitrogen (%)	Minimum 0.8	1.22	1.25	1.29
Total phosphate P_2O_5 (%)	Minimum 0.4	0.21	0.20	0.23
Potassium as K_2O (%)	Minimum 0.4	1.94	2.45	2.38
Organic carbon (%)	Minimum 12.0	23.18	25.72	24.80
C/N ratio	<20	19.0	20.6	19.22

Formatting Issue - Best Available Paper/Graphic

Table 3:

Effect of compost on crop yield

Crops	Yield in kg. per acre		Increase in yield (%)
	Control	With compost	
Paddy	2625 ± 55	3150 ± 50	20.0 ± 3.0
Amaranthus	7045 ± 150	9473 ± 120	34.5 ± 3.5
Bhendi	5406 ± 95	7082 ± 75	31.0 ± 3.0
Chilly	517 ± 20	618 ± 20	19.5 ± 4.0

Formatting Issue - Best Available Paper/Graphic

Figure legends

Fig. 1. Preparation of organic compost at pilot level (500 kg)

Fig. 2. Structural and molecular breakdown of keratin by keratinase

Fig.3. Scale up of hair degradation at 50 L and 500 L level in fermentor. (a) Degree of hydrolysis and keratinase production and (b) Profile of pH and dissolved oxygen during degradation

BIOLOGICAL LIQUEFACTION AND ANAEROBIC DIGESTION OF WASTE FLESHINGS INTEGRATED WITH SLUDGE AND BIO-ENERGY GENERATION – A NOVEL AND SUSTAINABLE DEVELOPMENT

SENGODA GOUNDER Rajamani[1], ARNOLD Mulder[2]
[1]Chairman, Asian International Union of Environment Commission (AIUE), INDIA.
[2]Amecon, Delft, The Netherlands

Corresponding Author: Dr.S.Rajamani,
E-mail: dr.s.rajamani@gmail.com

ABSTRACT

The safe disposal of large amount of solid wastes from tanneries such as unutilized fleshing and sludge from effluent treatment plants are major challenges and disposal of sludge in the Secure Land Fill (SLF) system is becoming prohibitive and environmentally challenging. The tanneries in Asia process 8 → 9 million tons of hides and skins per year and generate 3 to 4 million tons of solid waste and 2.5 to 3.5 million tons/year of dewatered sludge from effluent treatment plants. The disposal of sludge and solid waste especially about 1.0 million tons of fleshing are becoming a major environmental challenge.

The biological liquefaction of waste / unutilized fleshing in pilot scale by mixing with active bio-sludge / liquid from Anaerobic Digester could be achieved with a detention time of 7-8 days under atmosphere temperature of 20-35°C. Biologically liquefied fleshing is mixed with bio-sludge from effluent treatment plant with 4-6% solids concentration. Anaerobic Digestion is effective with a detention time of 30 days and generates 0.5m3 of bio-gas per kg of COD removal. Bio-gas is collected and stored in gas balloon and further converted into electricity using Gas Engine. The digested sludge is dewatered and converted into fertilizer by composting process.This unique and sustainable technological development will reduce the generation of hazardous category sludge and environmental issues in the disposal of unutilized fleshing.

Key words : Biological liquefaction, Environmental issue, Anaerobic digester, Fleshing from Tanneries, Fertilizer.

1. INTRODUCTION

The safe disposal of large amount of solid wastes from tanneries such as unutilized fleshing and sludge from effluent treatment plants are major challenges and disposal of sludge in the Secure Land Fill (SLF) system is becoming prohibitive and environmentally challenging. The tanneries in Asian countries process 8 → 9 million tons of hides and skins per year and generate 3 to 4 million tons of solid waste and 2.5 to 3.5 million tons/year of dewatered sludge from effluent treatment plants. The disposal of sludge and solid waste specifically waste fleshing are becoming a major environmental challenge. The tanneries processing raw hides and skins generate 10-15% of fleshing as wet solid waste. It is estimated annually on an average 1.0 million tons of fleshing are generated from Asian tanneries. The demand for conversion of fleshing into glue for wood works and other products are reducing due to the replacement by synthetic adhesives. Newenvironmental pollution control norms such as reduction in volatile organic matter in the sludge for disposal into Secured Land Fill (SLF) and lack of capacity requirement for establishment of organized unit for conversion into useful products, many small-scale unorganized units handling / processing waste fleshing vanished in many tannery clusters. Hence, it is becoming a necessary to develop and establish a viable environmental solution such as biological liquefaction, anaerobic digestion with bio-energy generation and bio-fertilizer to ensure safe disposal of waste fleshing and sludge.

2. METHODOLOGY OF BIOLOGICAL LIQUEFACTION

With a view to reduce the sludge generation from tannery effluent treatment plants and conversion into bio-energy and bio-fertilizer an anaerobic digestion system has been developed and applied on commercial scale under Indo-Dutch Technical Co-operation in the area of Environmental Protection to address the Leather industrial wastes. During the up-scaling of anaerobic digestion, an innovative development of biological liquefaction of waste fleshing by fermentation using part of bio-sludge from Anaerobic Digester was made and the liquefied fleshing successfully converted into bio-energy and bio-fertilizer.This innovation is separation of liquefaction process of the bio-mass mainly by enzymes and anaerobic digestion mainly by methane producing bacteria in different units. A part of active bio-mass from Anaerobic Digester is collected and slowly mixed with fleshing in serious of tanks for fermentation & liquefaction with a detention time of 7-10 days under atmosphere temperature of 18-35°C. The biologically liquefied fleshing becomes a fine slurry with high concentration of COD in the range of 20000 - 30000mg/l.

The process of biological liquefaction and anaerobic digestion is shown in the following process flow diagram.

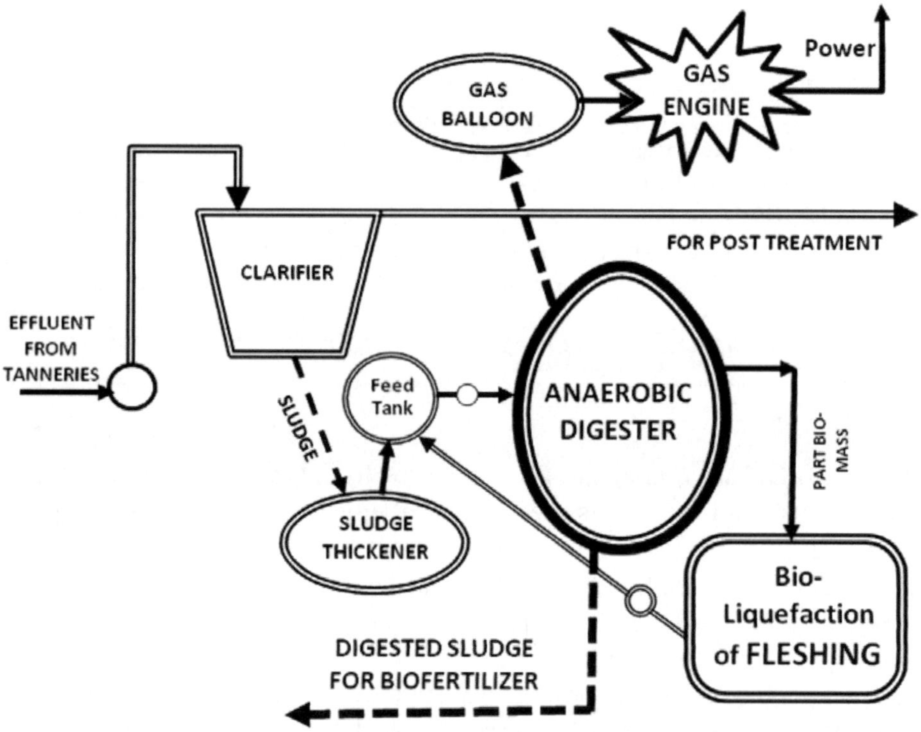

ANAEROBIC DIGESTION OF SLUDGE WITH LIQUEFIED FLESHING

The addition of Fermented Supernatant/Bio-mass from Anaerobic Digester, collection of waste fleshing from tanneries, making into small pieces / mincing by mechanical and mixing of fermented supernatant and fleshing in tanks are the main physical activities. Maintenance of detention time of 7-10 days for biological liquefaction by gentle intermittent stirring, collection of liquefied fleshing at the bottom of tanks and pumping the liquefied fleshing with high COD of 20000-30000 mg/l to the anaerobic reactor are shown in the following diagram.

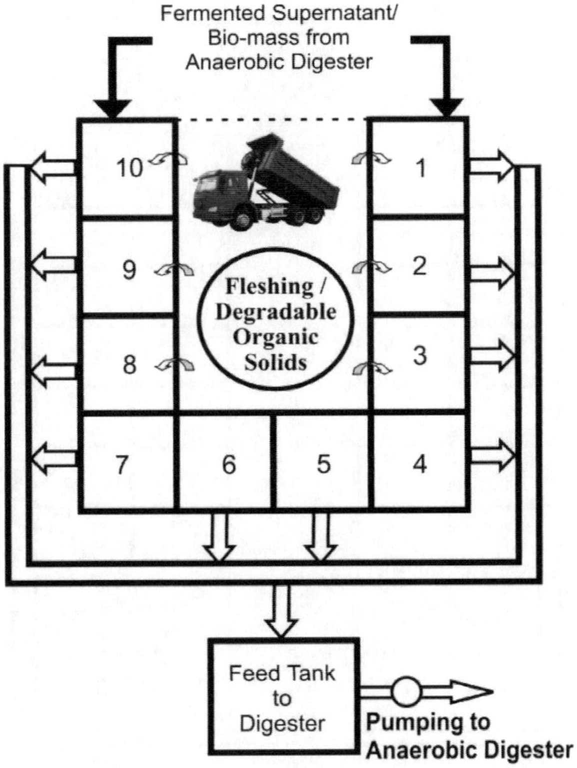

BIOLOGICAL LIQUEFACTION OF FLESHING

3. FEED CHARACTERISTICS AND ANAEROBIC REACTION

The liquefied fleshing is mixed with bio-sludge from effluent treatment plant with 3-6% solids and fed to the Anaerobic Reactor with COD load of 1900-2600kg/day. The characteristics of feed to Anaerobic Reactor are given in the following table.

Characteristics of Feed to Anaerobic Reactor/Digester

Sl. No.	Source of Feed	Volume	Solids Concentration	COD load in Kg
1.	From Biological Liquefaction System	5-10m3/day	3 – 5%	400-600
2.	Sludge from Primary & Secondary Clarifier of Soak Stream	20-25m3/day	4– 5%	800-1000
3.	Sludge from Primary & Secondary Clarifier of Composite Stream	15-20m3/day	4– 6%	700-1000
	Total feed / Rated capacity of Anaerobic Reactor	**Upto 55m3/day**	**3.5 – 5%**	**1900-2600**

During Anaerobic Digestion process with detention time of 30 days about 0.5m3 of bio-gas is generated per kg of COD removal. The bio-gas is collected and stored in gas balloon and further converted into electricity using Gas Engine.

The digested sludge is dewatered to increase the solids concentration to the level of 20-25% and taken for bio-fertilizer plant by mixing with degradable organic solid wastes such as vegetable market waste, leaves, etc., and converted into bio-fertilizer.The composting process is shown in the following process flow diagram.

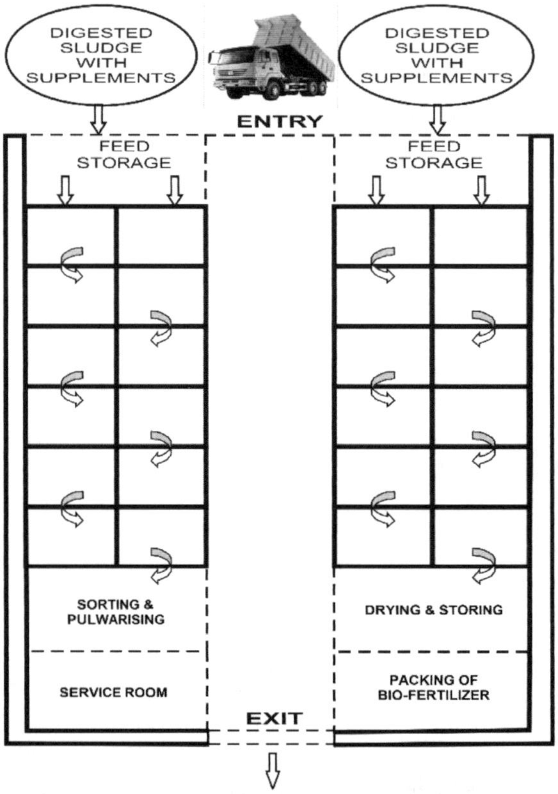

BIO-FERTILIZER FROM BIO-SLUDGE AND DEGRADABLE SOLID WASTE

4. RESULTS AND DISCUSSION

The rate of COD load, bio-gas generation per day and the corresponding electricity generation in kWh/day is given in the following table.

Bio-gas Production and Potential Electricity Generation from 50-55m^3/day

Sludge dry weight in %	Degradable dry weight load in kg/day	COD load in kg/day	Biogas production in m^3/day	Electricity generation in kWh/day
3	1500	1500	450	945
4	2000	2000	600	1260
5	2500	2500	750	1575
6	3000	3000	900	1890

- It is feasible to liquefy the waste fleshing biologically and increase the bio-mass feed to Anaerobic Reactor in terms of COD to more than 20000 mg/l.
- Optimum detention time for biological liquefaction is in the range of 7-10 days depending upon the fat content and atmospheric temperature.
- Optimum detention time for Anaerobic digestion is in the range of 25-30 days depending upon the COD load and atmospheric temperature.
- The bio-gas generation is in the range of 0.4-0.6m3 per kg of COD removal.
- Improved occupational health and safety measures in the context of COVID-19 pandemic situation had become a major threat not only for the technological development but also for sustainable operation and maintenance of the entire leather process and environmental management.

5. CONCLUSION

This unique and sustainable technological development of biological liquefaction of fleshing, anaerobic digestion by mixing with degradable sludge from effluent treatment plant and solid wastes and conversion into bio-fertilizer and energy will reduce the hazardous category sludge generation and minimize the disposal problem into SLF.The environmental issues in the disposal of unutilized fleshing and bio-sludge from effluent treatment plants are also minimized. Based on the pilot scale development, commercial scale anaerobic digesters are being implemented in Effluent Treatment Plants (ETPs) and CETPs in India, The Netherlands and other countries.

6. ACKNOWLEDGMENT

Contributions of Indian Leather Technology Association (ILTA),IULTCS, UNIDO, National Mission for Clean Ganga (NMCG), Department for Promotion of Industry and Internal Trade (DPIIT), National Green Tribunal (NGT), Mega Leather Cluster (MLC)-Kanpur, Schoolnet India Limited, Asian International Union Environment (AIUE) Commission, Asian International Forum and others commission members from various countries, AMECON & TNO Netherlands, Central Leather Research Institute (CSIR-CLRI), European Union including Italy, Spain, The Netherlands and other Countries such as China, Japan, Romania, Turkey, Taiwan and Russian Federation, New Zealand are acknowledged. Leather Industry Associations and Common Effluent Treatment Plants (CETP) specificallyPallavaram, Dindigul, Madhavaram, Jajmau, Banthar, Unnao CETPs in India are acknowledged.

7. REFERENCES

1. S.Rajamani (2020), *Innovative Ecological Processes With Recovery Of Chemicals And Water For Reuse In Leather Sector – An Economical And Sustainable Approach* – Article published in Journal of Indian Leather Technologists Association (JILTA), December 2020 issue.
2. S.Rajamani (2020), *Sustainable ZLD System by Adopting Centralized Treatment of Segregated Streams for Recovery of Reusable Quality Chemical, Salt & Water – First of its kind in India & Asia* – Article published in Leather News India Journal.
3. S.Rajamani (2020), *Sustainable Environmental Management in Indian & Asian Leather Industries*– Paper presented in XVI International Scientific-Practical Conference Leather and Fur In the XXI Century: Technology, Quality, Environmental Management, Education in Russian Federation - November 2020, Ulan Udi, Russia.
4. S.Rajamani (2019), *Innovative Technologies on Cleaner Production and Waste Management in Tanneries* – Paper presented in International Leather Engineering Congress Innovative Aspects for Leather Industry, Izmir, Turkey.
5. Arnold Mulder (2015), *Development of Cleaner Technology and Wastewater Treatment System* in *tanneries in Trujillo, Peru* – Senior Experts Mission report for for the foundation PUM Netherlands.
6. Shweta Singh (2015), *Reducing Waste in Indian Leather Industry to Suffice Profits*, Leather News India in the August 2015 edition.

Key technologies for tannery wastewater treatment

Wolfram Scholz[1]

W_2O Environment Ltd.

+ 43 676 782 9383

wolfram@w2oenvironment.net

Abstract

Tannery wastewater is a complex mixture of organic substances derived from the hide and inorganic substances such as salts and chemicals, which are added during the beamhouse and tanyard processing. Sulphide and chrome bearing streams have to be segregated well from the soak and de-lime effluents and are treated separately in a sulphide oxidation system and chrome precipitation plant. Sulphide oxidation is carried out in batch using blower assisted jetox aeration and achieves a complete oxidation of sulphides to sulphates within 8 hours. The segregated effluents are fine screened at 1 mm to remove coarse solids and mixed and aerated during the balancing stage to provide an uniform effluent for the following Primary treatment. Dissolved Air Flotation removes efficiently suspended solids, as well as fats, surfactants and colour of the wastewater achieving a clear effluent with < 50 mg/l Suspended Solids. The residual soluble COD and BOD is then treated biologically with a combination of anoxic and aerobic treatment to also allow removing nitrogen compounds such as ammonia, nitrate and Kjeldahl nitrogen. Biological denitrification/nitrification in combination with membrane bioreactor technology achieves a high reduction of 92% COD and 99% BOD, with ammonia of less than 2 mg/l achieved in the effluent. The membrane bioreactor permeate is solids free and can be polished with Nanofiltration to allow for up to 75% high quality water recycling, thus reducing the discharged volume to 25%. The Nanofiltration permeate is completely clear and contains only minor concentrations of salt and has a consistent quality enabling high quality water re-use.

---

[1] Gymnasiumstr. 6, Top 12, 1180 Vienna, Austria

Keywords: Tannery effluent, sulphide oxidation, nitrogen removal, MBR, membrane bioreactor, NF, nanofiltration, water recycling

Abbreviations

BOD	biological oxygen demand (mg BOD,L^{-1})
COD	chemical oxygen demand (mg COD,L^{-1})
DAF	dissolved air flotation
MBR	membrane bioreactor
MLSS	mixed liquor suspended solids (mg,L^{-1})
N	nitrogen
NF	nanofiltration
P	phosphorus
SS	suspended solids (mg,L^{-1})

Introduction

Most of the pollution generated during leather processing evolves from the beamhouse processes, which are the initial steps of leather production. The conventional beamhouse process involves soaking of hides, unhairing with sulphides and lime, following tanyard with de-liming with ammonium salts, degreasing, especially in ovine processing and pickling to prepare the hide for tanning. Within these processes organic substances as proteins, fats and hair, which account for approximately 80 % of the salted weight of the hide are removed and are being discharged from the process as highly organic loaded effluents (IUE 6, 2019).

The raising production and environmental costs urge the leather industry to implement sustainable production schemes with the main objectives of saving resources by process- and production-integrated measures to minimise environmental impacts. By strictly regarding economical demands, an approach of "green-processing" has been developed, that enables the leather industry to reduce environmental and operational costs. The implementation of processes-integrated measures and recycling technologies using Membrane bioreactor technology and Nanofiltration has shown to be technical and economical feasibility for recycling and to fulfil the quality criteria for leather processing.

Picture 1: Wet-blue tannery with effluent treatment plant discharging to a water course

Treatments

Segregation and screening of effluents

A process-controlled drainage from the drums is critical for successful wet-blue tannery effluent treatment to achieve a good segregation of chrome and sulphide bearing effluents without any cross contaminations (TAN, 2011). In wet-blue tanneries this is achieved by draining the lime float and tanning float into separate channels and collection in 3 individual sumps; the lime sump, the chrome sump and the sump for general effluents.

The 3 effluent streams are then fine screened with automated self-cleaning screens (IUE, 2021) to remove all particles > 1 mm. The implementation of fine screening reduces significantly the solids load for the following treatments and the overall treatment costs.

Sulphide Oxidation

The catalytic oxidation by aeration of lime liquors is the most economical and widely used process for sulphide removal. The technique consists of aerating the spent lime liquors for 8 hours. The sodium sulphide present in the spent un-hairing - liming liquors is oxidised by the air oxygen into thiosulphate, and in smaller quantities, into sulphate. The thiosulphate then decomposes into sulphur and sulphite. The amount of oxygen required is estimated at 2 kg per kg of sulphide (S^{2-}). However, the amount of air, which will react in relation to the quantity of air injected, will depend on the aeration device. Therefore, it is necessary to provide for an excess of air to the level of sulphide. Depending on the length of the float used and the volume of wash and rinse waters the lime liquors may contain between 1.5 to 3 g/l of sulphides. After mixing and dilution with the rest of the tannery effluent the sodium sulphide concentration is between 200 - 600 mg/l. Therefore, it is required to separately treat the unhairing-liming floats before they are mixed with other effluent streams. Sulphides can cause inhibitions and reduce the performance of the following biological treatment and have to be reduced to concentrations of <1 mg/l (as S^{2-}), whereas typical sulphide levels in the untreated effluent is of the order of several hundred.

The operation of a sulphide oxidation system is the most cost effective method of oxidising sulphides to sulphates offering advantages compared to other methods such as ferric salt addition which results in high generation of sludge or addition of expensive oxidants such as hydrogen peroxide with high chemical costs. A sulphide oxidation consists of a jetox venturi system connected to a centrifugal pump and an external blower to mix the lime liquors with air (IUE, 2021). Sulphide oxidation systems, operated in batch over 8 hours, have shown to achieve complete oxidation of lime liquors reaching residual levels of < 1 mg/l sulphides (IUE 12, 2019).

Balancing

Balancing of tannery effluent is important to improve the treatment efficiency, and avoid oversize treatment plants, which have to deal with peak effluent flows. A steady uniform effluent flow can be achieved by sizing the equalisation tank to a volume of 70 to 100% of the tannery daily flow. In addition to flow balancing, the equalisation tank provides for neutralisation and precipitation. It is necessary to provide effective mixing by mechanical stirring and air injection to achieve equalisation, prevent anaerobic conditions and settling of suspended solids. The installation of a jetox aerated balancing tank enables complete mixing and aeration of the effluents avoiding re-formation of sulphides (IUE, 2021).

Dissolved Air Flotation

Dissolved Air Flotation (DAF) works on the reverse principle to sedimentation, employing fine air or gas bubbles to lift suspended solids, which were not removed by the prior settlement, to the surface from where they may be removed. Especially fine solids as hair and fibres or fats and proteins can be very efficiently removed by flotation (TAN, 2011)

The flotation process relies on coagulant and flocculent chemical conditioning of the feed stream, as for sedimentation, in order to enhance the solids separation process. The effluent feed is pH adjusted, followed by dosing of a suitable coagulant. A suitable polyelectrolyte flocculant is also required for optimum phase separation, especially of colloidal solids, and will require inline dosing just prior to the effluent entering the flotation tank. After chemical dosing the effluent flows into the flotation tank and is mixed with the rising air bubbles stream. The air-saturated stream is formed by pumping of the treated effluent into a pressurisation chamber along with air, which under pressure, dissolves in the water. The sudden release of pressure in the flotation tank causes the dissolved air to form 'clouds' of tiny air bubbles, which come up to the surface carrying the suspended and colloidal solids with it to form a surface sludge 'blanket', which is regularly scraped off (IUE,

2021). A DAF system provides higher levels of pollution removal with up to 98% of SS and up to 75% of COD (IUE 5, 2019). A further advantage is that the floated sludge has a higher solids content, up to 10% dry matter, which is much easier and more cost effective to handle than just settling.

Picture 2: Primary effluent treatment plant with screening, balancing, sulphide oxidation and DAF

Biological treatment

Biological processes are based on the fact that micro-organisms can use the organics (BOD) and nutrients (N&P) to build up their cell mass as well as provide their energy requirements. The soluble organics are converted to CO_2 and wasted sludge from the biomass. The biological stage of a waste treatment plant can be either operated under aerobic or anoxic conditions. Aerobic conditions are applied to remove BOD and ammonia of tannery wastewaters, while anoxic conditions are used to convert nitrates to nitrogen in the process of de-nitrification. The activated sludge process is the most common biological treatment method for tannery wastewater treatment. Prolonged aeration time is important for tannery effluents due to the high organic load. De-nitrification loops have been added to biological treatment plants, where low ammoniacal nitrogen discharge limits are imposed (IUE, 2021). The denitrification/ nitrification systems are operated in sequence with effluents recirculating in between the aerated zone and the anoxic zone have shown to consistently reduce ammonia to < 2 mg/l and nitrate to < 5 mg/l (Scholz, 2013).

Membrane Bioreactor (MBR)

Membrane bioreactors (MBRs) combine an activated sludge process with membrane ultrafiltration to facilitate complete retention of the biomass. The process relies on membrane filtration to effectively retain all the biomass in the bioreactor as opposed to conventional treatment, where the biomass is continuously lost during clarification. As a consequence, the MBR process is operated at 3 folds higher mixed liquor concentrations of up to 10 g/l MLSS, than conventional biological treatment. This increase of concentration of wastewater bacteria allows reducing the aerated biological tank dimensions by 55%, compared to conventional biological treatment, which consequently also reduces aeration requirements by 35% (IUE, 2021).

A further benefit of MBR, resulting from increased sludge retention times and operating temperatures, is the reduction of surplus sludge generation. Due to the high sludge retention time and high MLSS, the bioreactor is severely carbon limited and the influent carbon sources are predominantly utilized for metabolic purposes than other cellular growth, e.g., maintenance, thus minimizing biomass production. Large scale MBR plants have shown that only 6.7% of the metabolized COD was incorporated into surplus sludge production, compared to 30-50% in a conventional activated sludge system. This minimizes the surplus sludge and consequently the amount of disposable biological sludge (Scholz, et.al, 2005).

Membrane bioreactor technology is an excellent pre-treatment for nanofiltration or reverse osmosis plants due to the complete elimination of BOD and suspended solids. Small concentrations of BOD after conventional treatment would cause bio-fouling of NF/RO membranes and suspended solids would physically block the spacer of a spiral membrane module. The combination of a Nanofiltration plant applied as a polishing step after MBR treatment allows high quality water recycling back into the tannery process. The advantage of this is a reduction of fresh water and effluent costs. The use of consistent high quality process water will also have a positive influence in regards of saving process chemicals and improving the leather quality. A membrane bioreactor system achieves up to 95% COD, 99% BOD removal and suspended solids free and transparent permeate (IUE 5, 2019).

Picture 3: Biological effluent treatment plant with anoxic and aerobic biological MBR treatment

Nanofiltration

The application of recycling normally relies on suitable process technology for water purification. The wide fluctuation in tannery effluent quality coupled with the requirements for process water of reliable quality tend to favour the application of membrane processes. Membrane filtration processes inevitably play nowadays a key role in modern water recycling since they can produce a water of consistent and reliably high quality. Membranes form a highly selective barrier and are tolerant to shock-loads. Therefore the produced permeate quality is varying little with the feed water quality (Judd S. and Bennett, 2004). The main advantage of a membrane-based process is that the concentration and separation is achieved without a change of state and without use of chemicals or thermal energy, thus making the process energy efficient and ideally suitable for recycling (IUE, 2021).

Nanofiltration technology is a suitable, where a final polishing of effluents, without complete salt retention is required. The Nanofiltration membrane offers a small pore size of 400 – 600 Dalton, which retains efficiently multivalent ion such as total hardness and certain charged or polar molecules. However, sodium chloride a mono-valent salt passes the membrane. The spiral-wound modules are densely packed offering a high membrane surface and therefore require only minimum of space. High reductions of COD, BOD and colour are achieved, due to the fact that the Nanofiltration membrane retains organic fractions. The produced permeate is reduced in COD, completely clear and contains minor concentrations of salt (Scholz, 2017). The consistent quality of the NF enables high quality water.

Results

The segregation and fine-screening of sulphide and chrome bearing effluents and separate treatment with sulphide oxidation and chrome precipitation is essential for wet-blue tannery effluent treatment.

The sulphide oxidation using blower assisted jetox venturi aeration has shown to remove efficiently sulphides with average concentration of 0.19 ppm achieved with the completion of an 8 hr oxidation cycle.

Dissolved Air Flotation has shown to remove 75 % of COD, achieving a clear and transparent effluent with < 50 mg/l SS. DAF treatment achieves a high Primary sludge dryness of up to 10% and therefore reduces considerably the volumes of sludge for de-watering.

Biological treatment with anoxic pre-denitrification followed by aerobic treatment has shown to reduce ammonia to < 2 mg/l and total nitrogen to < 5 mg/l.

The activated sludge is then filtered with submerged membranes resulting in a membrane bioreactor permeate, achieving 92% COD removal with a residual BOD of < 20 mg/l.

Nanofiltration has shown to remove residual organics and colour, achieving water recovery rates of up to 75% of a clear and soft effluent, which is ideal for process water recycling.

Conclusions

The combination of Membrane Bioreactors (MBR) with Nanofiltration (NF) for water recycling is a new strategy to reduce the environmental impact of tannery, reaching steadily the stringent discharge compliance and to provide for high quality process water recycling. With proven economical and technical feasibility, this novel technology has shown reductions of 92 % in COD, 98 % in BOD and complete removal of suspended solids. The high-quality effluent after MBR treatment, allows for cost-effective polishing through NF treatment and high quality water recycling. Full-scale applications of MBR and NF treatment have demonstrated 75 % water recovery.

References

[1] IUE Tannery effluent treatment videos: Animations of treatment technologies (2021): https://iultcs.org/tannery-effluent-treatment-videos/

[2] Environmental Document of the IULTCS. IUE 5 - Typical performance for tannery wastewater treatment (2019): https://iultcs.org/wp-content/uploads/2020/07/IUE_5.pdf

[3]Environmental Document of the IULTCS, IUE 6 - Typical pollution values related to conventional tannery processes, (2019): https://iultcs.org/wp-content/uploads/2020/07/IUE_6.pdf

[4] Environmental Document of the IULTCS, IUE 12 (2019): https://iultcs.org/wp-content/uploads/2020/07/IUE_12.pdf

[5] TAN (2011) - Best Available Techniques for the Tanning of Hides and Skins, chapter 4 p. 183

[6] Scholz W., (2013): Membrane bioreactor treatment of tannery effluents with nitrogen removal and low cost sludge drying. Proceedings for the XXXII Congress of the IULTCS, Istanbul, Turkey

[7] Scholz W., Rougé P., Bodalo A. and Leitz U. (2005) Desalination of mixed tannery effluent with Membrane Bioreactor and Reverse Osmosis Treatment. Journal of Environmental Science & Technology. p. 39:21, 8505 – 8511

[8] Judd S. and Bennett, (2004): A. Membrane Technology recovers water for industrial reuse. Industrial water and wastewater treatment, p.13-14

[9] Scholz W., (2017) Clean Salt Recovery using Nanofiltration. Proceedings for the XXXIV Congress of the IULTCS. Chennai, India

Analysis of Post-tanning Chemicals and Reduction of Pollution Load in Wastewater

Éverton Hansen[1,2,3,4*], Patrice Monteiro de Aquim[2], Mariliz Gutterres[1]

[1]*Federal University of Rio Grande do Sul - UFRGS, Post-graduation Program of Chemical Engineering, Laboratory for Leather and Environmental Studies -LACOURO, Eng. Luiz Englert street, Porto Alegre, RS, Brazil*

[2] *Feevale University, Institute of Pure Sciences and Technology, 2755 RS 239, Novo Hamburgo, RS, Brazil*

[3]*Centro Universitário Ritter dos Reis – UniRitter, 555 Orfanotrófio, Porto Alegre – RS, Brazil.*

[4]*Universidade do Vale do Rio dos Sinos, Chemical Engineering Department, 950 Av. Unisinos, São Leopoldo, RS, Brazil*

* Corresponding author email address: evertonhansen@gmail.com

ABSTRACT:

The chemicals in post-tanning are partially retained by the leather, and the fractions remaining in the process are discharged with the wastewater. Current literature has an extensive number of studies characterizing tannery effluents. However, few studies have evaluated how groups of chemicals applied to the leather impact the quality of wastewater generated in the process. Thus, this study aims to evaluate and reduce the impact of chemicals used in the post-tanning process on the main quality parameters of the wastewater. The assessment and reduction of the pollution load of leather chemicals were carried out using a formulation from a Brazilian tannery as a case study. Results showed that retanning agents (natural and synthetic tannins) were responsible for high inorganic pollution load in the wastewater, and the synthetic tannins were more toxic than the natural ones. The fatliquoring agents were responsible for the highest chemical oxygen demand load in the wastewater and were the chemical group that presented the highest cytotoxicity. The fixing agent and the dye contributed to the inorganic pollution load of the wastewater, and the nitrogen load of the wastewater was mainly related to the neutralizing retanner and the dye. The adjustment of the chemicals supply showed depletion in the pollution load of wastewater. The obtained leather showed quality within the required specifications of the tannery, and the formulation cost was reduced by 24%.

Keywords: leather; post-tanning; wastewater; dyes; fatliquoring agents; retanning agents; cytotoxicity.

1. Introduction

The leather industry and its products play a relevant role in the world economy, with a global trade value of approximately 80 billion dollars per year (Sivaram and Barik, 2019). This industry is especially important in the economy of developing countries (Bharagava and Mishra, 2018). According to the Center for Tanning Industries in Brazil - CICB (2020), the Brazilian tanning industry has 310 tanneries and 2,800 leather and shoe components factories. The Brazilian leather industry was responsible for exporting more than 1.1 billion dollars in 2019. Finished leather accounts for 59.6% of sales (ABQTIC, 2020).

In the post-tanning (wet-finishing) process the chemicals are not fully retained by the leather and are partially discarded in the raw wastewater of the tannery (Aquim et al., 2009; Bharagava et al., 2018; Hansen et al., 2021; Moreira et al., 2019). Some of these chemicals are difficult to remove in wastewater treatment plants, especially when they contain salts and recalcitrant compounds (Kozik et al., 2019; De la Luz-Pedro et al., 2019; Moreira et al., 2019; Tamersit et al., 2018). Furthermore, concepts of cleaner production and pollution prevention (EPA, 1988) identify the need to reduce or eliminate waste in volume, concentration, and/or toxicity at the source, applying a preventive and integrated environmental strategy to industrial processes. Pollution prevention measures are beneficial because they reduce operating costs, reduce ecological damage resulting from the extraction and processing operations of the raw materials used, reduce employees' exposure to harmful substances and improve the company's image (El-Haggar, 2007).

Some studies have already evaluated the chemical and toxicological characteristics of post-tanning chemicals (Libralato et al., 2011; Lofrano et al., 2007, 2008; Moreira et al., 2019). However, studies relating the characteristics of leather chemicals with their impacts on raw tannery effluents and evaluating the supply of leather chemicals in processing floats are still necessary to expand knowledge in this area. Thus, this study aimed to evaluate and reduce the impact of chemicals used in the post-tanning process on the physicochemical and cytotoxic parameters of the wastewater.

2. Materials and Methods

This research used as a case study a post-tanning formulation processing wet-blue bovine leather for producing women bag. This formulation is used by a tannery located in southern Brazil. The tannery has a production capacity of 750 whole leathers per day.

2.1 Characterization of Leather Chemicals

The leather chemicals offered in the process were analyzed individually at the concentration indicated in the formulation (Table 1) and characterized for conductivity

(Hach hq40d multi-analyzer), total dissolved solids (TDS) (Hach hq40d multi-analyzer), chemical oxygen demand (COD) (SM 5220 C, 2017), biochemical oxygen demand (BOD) (SM 5210 D, 2017), total kjeldahl nitrogen (TKN) (SM 4500 Norg, 2017), ammoniacal nitrogen (NH4-N) (SM 4500 NH3 C, 2017), and cytotoxicity. Cytotoxicity assays were performed using V 79-4 cells. The cells were exposed to a culture medium that was diluted in samples containing the chemicals for 24 hours (for acute cytotoxicity) and for 96 hours (for chronic cytotoxicity). At the end of this period, two independent endpoints were performed: tetrazolium reduction assay (MTT) (Mosmann, 1983) and Sulphorhodamine B adsorption (SRB) (Skehan et al., 1990).

Table 1: Concentration of chemicals added in post-tanning processing (Hansen et al., 2020).

Leather Chemical	**Concentration (g/L)**
Sodium bicarbonate	1
Sodium formate	10
Neutralizing retanner	10
Vegetable tannin 1	30
Vegetable tannin 2	10
Synthetic tannin 1	10
Synthetic tannin 2	20
Synthetic tannin 3	15
Acrylic resin	15
Natural oil 1	15
Natural oil 2	2,5
Mixture of synthetic and natural oils 1	15
Mixture of synthetic and natural oils 2	15
Dye auxiliary	2,5
Waterproofing agent	1
Fixing agent	7
Antioxidant	0,5
Fungicide	1
Black acid dye	12,5

2.2 Reducing Pollution Load of Chemicals in Wastewater

Tests to reduce the supply of chemicals were carried out in pilot-scale and in industrial-scale drums. Since retanning and fatliquoring agents are the chemicals used in greater quantity, the reduction in the supply of leather chemicals focused on reducing the consumption of these agents. The reduction in the offer of leather chemicals was of 26%, based on tests carried out previously (Hansen et al., 2021).

The residual wastewater from the pilot scale tests were characterized before and after the reduction in the supply of chemicals, evaluating pH, conductivity, and total dissolved solids using Hach hq40d multi-analyzer. The leather produced on a pilot scale was tested for the organoleptic parameters of softness, filling, frain, and impregnation hability.

The wastewater before and after the reduction in chemical supply on the industrial scale were analyzed for pH (SM 4500 H+, 2017), conductivity (SM 2510 B, 2017), TDS (SM 2540 C, 2017), biochemical oxygen demand (BOD) (SM 5210 D, 2017), COD (SM 5220 C, 2017), and sulfate (SM 4110 B, 2017). The leather obtained was tested for tear resistance strength (ISO 3377-1, 2011), colorfastness to light ASTM D1148 (2018), colorfastness to friction (wool felt wet with water and wool felt wet with alcohol) ISO 11640 (2012).

3. Results and Discussion

3.1 Characterization of Leather Chemicals

The characterization of each leather chemical of the post-tanning formulation is shown in Figure 1.

Figure 1: Conductivity (a), TDS (b), COD (c), BOD (d), TKN (e), and NH4-N (f) of post-tanning chemicals.

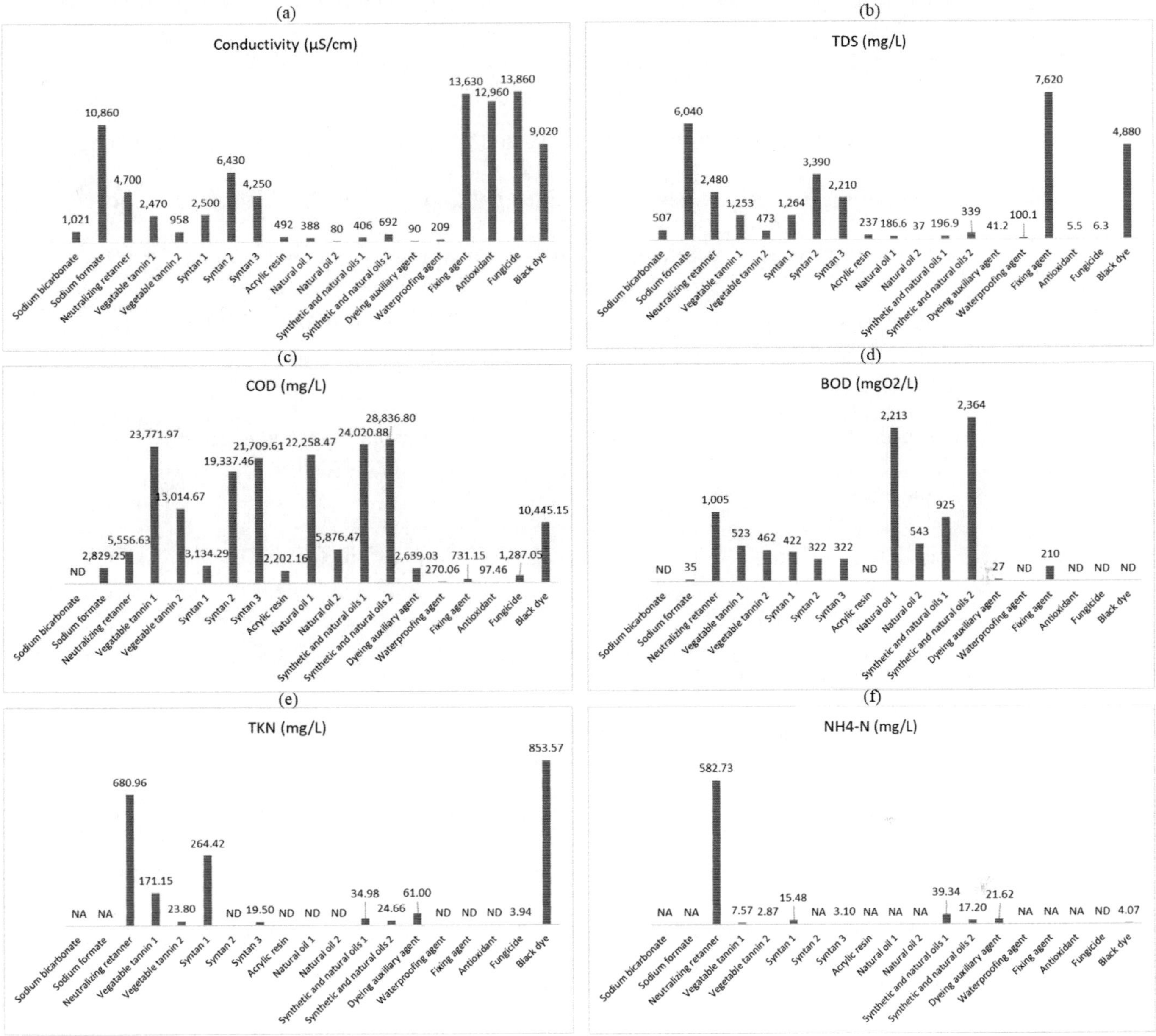

NA - not analyzed.
ND - not detected.

The analysis of the leather chemicals provided information on how each group of chemicals used in the formulation contributes to the pollution load of the effluents:

- Deacidulants showed high conductivity and TDS, and the neutralizing retanner showed the highest ammoniacal nitrogen pollution load in the formulation. The safety data sheet for this chemical indicates that it is an aromatic sulfone, which possibly has amine groups attached to its chemical structure, given the result of nitrogen.
- Retanning agents were responsible for high inorganic pollution load (conductivity and TDS).
- Fatliquoring agents were the chemical group responsible for the highest COD load.
- The fixing agent and the acid dye showed high conductivity and TDS and the acid dye also showed high total nitrogen pollution load.

Cytotoxicity results of the chemicals are shown in Table 2. The results are presented as cell viability compared to the negative control (control test without the presence of chemicals). Thus, the higher the cell viability percentage, the lower the toxicity of the chemical. Synthetic tannins were more toxic than vegetable tannins. Fatliquoring agents were the chemical group with the highest cytotoxicity, which was related in this study to the use of surfactants to obtain these chemicals, since oils are applied in floats as aqueous emulsions. The cytotoxicity of surfactants occurs due to the lysis of the cell membrane (Hansen et al., 2020). Fungicide also showed acute and chronic cytotoxicity, which was expected due to the presence of substances such as chlorobenzene in its composition.

Table 2: Cytotoxicity assays of chemicals.

	MTT 24h		MTT 96h		SRB 24h		SRB 96h	
Chemical	Average	Desvpad	Average	Desvpad	Average	Desvpad	Average	Desvpad
Sodium bicarbonate	99,3	8,3	75,2	3,2	102,0	1,3	89,5	9,8
Sodium formate	39,6	5,2	47,0	3,9	56,4	7,9	27,7	4,3
Neutralizing retanner	72,5	5,7	4,4	0,2	76,7	3,4	4,9	1,0
Vegetable tannin 1	40,3	2,3	51,3	8,0	87,7	3,1	42,9	0,9
Vegetable tannin 2	85,7	6,5	30,8	2,7	90,0	1,4	48,2	2,5
Synthetic tannin 1	43,7	5,0	5,3	0,5	41,2	4,6	1,7	0,4
Synthetic tannin 2	3,2	1,0	4,5	0,6	1,0	1,4	1,1	0,3
Synthetic tannin 3	4,5	4,5	4,7	0,9	0,6	0,4	0,4	0,2
Acrylic resin	18,8	5,2	5,4	1,2	23,1	5,9	1,7	0,6
Natural oil 1	3,3	1,1	0,0	0,1	4,7	1,2	3,4	0,6
Natural oil 2	8,6	6,4	3,2	1,1	4,2	0,6	1,1	0,1
Mixture of synthetic and natural oils 1	2,4	0,5	3,0	0,3	6,1	11,1	1,0	0,2
Mixture of synthetic and natural oils 2	0,7	1,3	5,8	0,8	4,2	3,4	2,4	0,1
Dyeing auxiliary agent	2,7	0,7	4,8	0,4	5,6	1,3	2,7	0,3
Waterproofing agent	121,6	7,2	11,4	2,8	87,9	3,3	12,6	4,9
Fixing agent	109,1	16,0	68,3	6,1	100,9	2,1	97,7	1,2
Antioxidant	111,3	5,5	78,2	2,1	105,3	0,6	99,1	1,7
Fungicide	3,3	0,7	4,5	0,5	4,5	1,7	4,8	1,1

3.2 Reducing Pollution Load of Chemicals in Wastewater

The adjustment in chemicals supply on a pilot scale resulted in reduction in conductivity and total dissolved solids in the wastewater (Table 3). The results of the analysis of the wastewater in industrial scale tests also showed a reduction in the concentrations of conductivity, TDS, BOD, and COD (Table 4). There was high exhaustion of the chemicals due to the absorption on the wet-blue leather, reaching up to 89.8% reduction of BOD, which is related to an improvement in the diffusion of oils, since the various post-tanning chemicals are less concentrated together in the float.

Table 3: Wastewater analysis on a pilot scale.

Residual float	pH	Condutivity (μS/cm)	TDS (mg/L)
Residual float I* (Original Process)	3.90	1,994	11,580
Residual float I (Chemical reduction)	3.75	1,789	9,440
Residual float II** (Original Process)	3.72	689	3.62
Residual float II (Chemical reduction)	3.62	586	3.10

*Residual Float I: retanning, fatliquoring, and dyeing wastewater
**Residual Float II: fatliquoring and fixing wastewater

Table 4: Wastewater analysis on an industrial scale.

	pH	Conductivity (μS/cm)	TDS (mg/L)	BOD (mg/L)	COD (mg/L)	Sulfate (mg/L)
Residual float I (Original process)	3,4	>5000	27,716.5	1,610	9,295.03	11,599.65
Residual float I (Chemical reduction)	3,4	>5000	23,633.0	1,258	6,710.13	10,033.96
Residual float II (Original process)	3,2	>5000	7,846.5	2,012	13,725.26	2,352.13
Residual float II (Chemical reduction)	3,2	>5000	6,211.0	205	7,292.37	1,885.61

The organoleptic evaluation of the leather showed that softness, filling and grain had equal mean results before and after chemical reduction on a pilot scale (Table 5), and the visual analysis of impregnation showed superior results after the reduction in the supply of chemicals, which may be related to the lower concentration of other chemicals, which would compete in the diffusion of the resin into the leather.

Table 5: Organoleptic evaluation of the leather.

	Original formulation		Chemical reduction	
	Average*	Desvpad	Average*	Desvpad
Softness	3	0	3	0
Filling	4	0	4	0
Grain	2.33	0.47	2.33	0.47
Impregnation	3	0	5	0

*scale from 1 (worst result) to 5 (best result)

The Physical-mechanical quality requirements were met on the leather produced on the industrial scale for all the tested parameters (Table 6). In addition to the reduction in the pollution load of the raw wastewater, there was a reduction of at least 24% in the process costs, considering only costs related to chemicals used in the formulation.

Table 6: Physical-mechanical tests on finished leather produced with chemical reduction (Hansen et al., 2021).

Physical-mechanical test	Result	Requirement
Tear resistance strenth (N)	107.91	60 for large handbag (ISO 3377-1:2011, 2011)
Colorfastness to light	5	≥ 3* (Basf, 2005)
Colorfastness to friction - water	5	≥ 3* (Basf, 2005)
Colorfastness to friction - alcohol	4 a 5	≥ 3* (Basf, 2005)

Conclusions

This study showed that the leather chemicals used in the formulation represent a relevant contribution to the organic and inorganic pollution load of raw wastewater. The results, both on pilot and industrial scale, showed that it was possible to reduce the pollution load of the raw wastewater, maintaining the organoleptic properties of the leather and meeting the established physical-mechanical requirements. In addition, formulation costs were reduced by 24%.

These results motivate more efforts to reduce the environmental impacts of tanneries, as the extrapolation of the use of this technique in other tanneries can reduce the mass of leather chemicals used and the pollution load of this industry's wastewater.

References

AQUIM, P. M. DE; GUTTERRES, M.; TRIERWEILER, J. Assessment of Water Management in Tanneries. v. 94, p. 253–258, 2009.

ASSOCIAÇÃO BRASILEIRA DOS QUÍMICOS E TÉCNICOS DA INDÚSTRIA DO COURO (ABQTIC). Guia Brasileiro do Couro - Brazilian Leather Guide. Estância Velha, 2020.

ASTM D1148. Test Method for Rubber DeteriorationDiscoloration from Ultraviolet (UV) or UV/Visible Radiation and Heat Exposure of Light-Colored Surfaces. 2018.

BASF. Vade-mécum do curtidor. Germany: BASF, 2005.

BHARAGAVA, R. N.; MISHRA, S. Hexavalent chromium reduction potential of Cellulosimicrobium sp. isolated from common effluent treatment plant of tannery industries. Ecotoxicology and Environmental Safety, v. 147, p. 102–109, 2018.

BHARAGAVA, R. N.; SAXENA, G.; MULLA, S. I.; PATEL, D. K. Characterization and Identification of Recalcitrant Organic Pollutants (ROPs) in Tannery Wastewater and Its Phytotoxicity Evaluation for Environmental Safety. Archives of Environmental Contamination and Toxicology, v. 75, n. 2, p. 259–272, 2018.

CICB, C. DAS I. DE C. DO B.-. O Couro e o Curtume Brasileiro. , 2020. Disponível em: <http://www.cicb.org.br/cicb/sobre-couro>.

EL-HAGGAR, S. M. Sustainable Industrial Design and Waste Management: Cradle-to-cradle for Sustainable Development. Academic Press, 2007.

EPA, E. P. A. Principles of Pollution Prevention and Cleaner Production. 1988.

HANSEN, É.; DE AQUIM, P. M.; GUTTERRES, M. Environmental assessment of water, chemicals and effluents in leather post-tanning process: A review. Environmental Impact Assessment Review, v. 89, p. 10, 2021.

HANSEN, É.; CARDOSO, J. K.; GUTTERRES, M.; MONTEIRO DE AQUIM, P. SCALE-UP TESTING FOR REDUCING POLLUTION LOAD OF CHEMICALS IN WASTEWATER OF LEATHER POST-TANNING. Process Safety and Environmental Protection, p. S0957582021005140, 2021.

HANSEN, É.; MONTEIRO DE AQUIM, P.; HANSEN, A. W.; et al. Impact of post-tanning chemicals on the pollution load of tannery wastewater. Journal of Environmental Management, v. 269, p. 110787, 2020.

ISO 3377-1:2011. Leather — Physical and mechanical tests — Determination of tear load Part 1: Single edge tear. 2011.

ISO 11640. Leather — Tests for colour fastness — Colour fastness to cycles of to-and-fro rubbing. 2012.

KOZIK, V.; BARBUSINSKI, K.; THOMAS, M.; et al. Taguchi Method and Response Surface Methodology in the Treatment of Highly Contaminated Tannery Wastewater Using Commercial Potassium Ferrate. Materials, v. 12, n. 22, p. 3784, 2019.

LIBRALATO, G.; AVEZZÙ, F.; VOLPI GHIRARDINI, A. Lignin and tannin toxicity to Phaeodactylum tricornutum (Bohlin). Journal of Hazardous Materials, v. 194, p. 435–439, 2011.

LOFRANO, G.; AYDIN, E.; RUSSO, F.; et al. Characterization, Fluxes and Toxicity of Leather Tanning Bath Chemicals in a Large Tanning District Area (IT). Water, Air, & Soil Pollution: Focus, v. 8, n. 5–6, p. 529–542, 2008.

LOFRANO, G.; MERIÇ, S.; BELGIORNO, V.; NAPOLI, R. M. A. Fenton's oxidation of various-based tanning materials. Desalination, v. 211, n. 1–3, p. 10–21, 2007.

DE LA LUZ-PEDRO, A.; MARTÍNEZ PRIOR, E. F.; LÓPEZ-ARAIZA, M. H.; et al. Pollutant Removal from Wastewater at Different Stages of the Tanning Process by Electrocoagulation. Journal of Chemistry, v. 2019, p. 1–9, 2019.

MOREIRA, M.; HANSEN, E.; GIACOMOLLI, G.; MORISSO, F.; AQUIM, P. Evaluation of Chemical Products in Leather Post-tanning Process and Their Influence in Presence of Neutral Salts in Raw Tannery Effluent. v. 114(4), p. 108–117, 2019.

MOSMANN, T. Rapid colorimetric assay for cellular growth and survival: application to proliferation and cytotoxicity assays. v. 65, p. 55–63, 1983.

SIVARAM, N. M.; BARIK, D. Toxic Waste From Leather Industries. Energy from Toxic Organic Waste for Heat and Power Generation. p.55–57, 2019. Woodhead Publishing.

SKEHAN, P.; STORENG, R.; SCUDIERO, D.; et al. New colorimetric cytotoxicity assay for anticancer-drug screening. , v. 82, p. 1107–1112, 1990.

SM 2510 B. Standard Methods for the Examination of Water and Wasterwater. Method 2510 B - Conductivity - Laboratory Method. 23º ed. 2017.

SM 2540 C. Standard Methods for the Examination of Water and Wasterwater. Method 2540 C - Solids in Water. 23º ed. 2017.

SM 4110 B. Standard Methods for the Examination of Water and Wasterwater. Method 4110 B - Ion Chromatography with Chemical Suppression of Eluent Conductivity. 23º ed. 2017.

SM 4500 H+. Standard Methods for the Examination of Water and Wasterwater. Method 4500 H+ - pH Value in Water by Potentiometry Using a Standard Hydrogen Electrode. 23º ed. 2017.

SM 4500 NH3 C. Standard Methods for the Examination of Water and Wasterwater. Method 4500 NH3 C - Titrimetric Method. 23º ed. 2017.

SM 4500 NORG. Standard Methods for the Examination of Water and Wasterwater. Method 4500 Norg - Semi-Micro-Kjeldahl. 23º ed. 2017.

SM 5210 D. Standard Methods for the Examination of Water and Wasterwater. Method 5210 D - Respirometric Method. 23º ed. 2017.

SM 5220 C. Standard Methods for the Examination of Water and Wasterwater. Method 5220 C - Closed Reflux, Titrimetric Method. 23º ed. 2017.

TAMERSIT, S.; BOUHIDEL, K.-E.; ZIDANI, Z. Investigation of electrodialysis anti-fouling configuration for desalting and treating tannery unhairing wastewater: Feasibility of by-products recovery and water recycling. Journal of Environmental Management, v. 207, p. 334–340, 2018.

New frontiers of Advanced Diagnostics and Non-Destructive Testing for quality control in the tanning industry

Florio C.[1], Pagliarulo V.[2], Leone G.[2], d'Angelo G.[2], Renò V.[3], Attolico G.[3], Stella E.[3], Ferraro P.[2]

1) *SSIP - Stazione Sperimentale per l'Industria delle Pelli e delle materie concianti - Italian Leather Research Institute - Via Campi Flegrei 34 Pozzuoli, Italy*
2) *ISASI - CNR (National Research Council) Via Campi Flegrei 34 Pozzuoli, Italy*
3) *STIIMA - CNR (National Research Council), via Amendola, 122/D, Bari, Italy*

Corresponding author: c.florio@ssip.it

Abstract. Leather is one of the most representative materials of the leading sectors of world markets, such as fashion and luxury, interior design, automotive, thanks to its excellent properties able to guarantee its high quality.

Although material of natural origin, leather undergoes numerous substantial transformations during the production process, which, together with its anisotropy and with the increasing use of a wide range of novel sustainable tanning systems, sometimes makes some defects unpredictable.

This work describes how some specific instrumental detection techniques are able to provide information regarding the possible causes of defects, also revealing the specific phase of the tanning process potentially responsible for the problem; particularly, the study emphasizes the effectiveness of techniques such as ATR-IR (Attenuated Total Reflection Infrared) Spectroscopy, Scanning Electron Microscope (SEM)-X-Ray Probe, Gas Chromatography-Mass Spectrometry (GC-MS), DSC/TGA (Differential Scanning Calorimetry/ Thermogravimetric Analysis) in identifying the causes of stains, chromatic or morphological alterations, cracks and detachment of the finishing.

Furthermore, in order to guarantee increasingly predictive and non-invasive approaches, Non-Destructive Testing (NDT) systems have been considered for their use in the quality control of tanning productions, with particular reference to the use of optical Interferometric Techniques, laser based, such as the Electronic Speckle Pattern Interferometry (ESPI), and of Computer Vision techniques, based on the acquisition and processing of signals and images at different wavelengths, from visible to infrared.

The possibility of using robotic systems specifically designed for the scanning of online products, through non-destructive and automated approaches, was also evaluated thanks to the increasingly widespread use of deep learning approaches, where these technological developments will certainly benefit the production chain to achieve ever higher standards of quality of leather products.

Keywords: ***Leather Defects; Advanced Diagnostics; Non-Destructive Testing; Quality Control; Novel Detection Techniques; Automated Approaches; Robotic Systems; Deep Learning.***

1 Introduction

There are numerous defects that can affect leather articles, and that can concern purely aesthetic aspects or can affect structural and performance features; sometimes, the causes underlying them can be complex, multiple, and simultaneously capable of compete to the onset of the defect. It should also be added that the progressive use of innovative tanning systems with low environmental impact, and with a consequent radical change in the production process, currently makes it more difficult to predict the type of defects that may affect the final articles.

On the other hand, the use of appropriate investigation tools can allow identifying the elements able to determine the problem, such as possible errors in the individual phases of the tanning process, rather than storage or assembly errors of the leather items.

Early diagnosis can also avoid possible negative repercussions for producers in economic and image terms, especially due to the possibility of onset of "hidden defects", macroscopically visible even after a long time and, in some cases, when the article it is already commercialized.

This work firstly provides an overview of the use of the main destructive and non-destructive diagnostic techniques, which have been found to be crucial for the analysis of a wide range of cases of defects on leather and leather items, as well as for the identification of the causes determining them, with the possibility of providing suitable activities for the prevention and minimization of critical issues.

The work also provides an overview of additional non-destructive and automated control tools, which can be effectively used for the early identification of defects[1,9], and definitely to ensure the quality of new generation leather products.

The main diagnostic techniques used in the first part of the work were:

- Attenuated Total Reflection Infrared Spectrometery (ATR-IR)
- Optical microscopy (OM)/ Phase-contrast microscopy (PCM)
- Scanning Electron Microscope equipped with X-Ray Probe (SEM-X-Ray Probe)
- Differential Scanning Calorimetry/ Thermogravimetric Analysis (DSC/TGA)
- Gas chromatography-mass spectrometry (GC-MS) equipped with autosampler for VOC analysis

<u>ATR-IR (Attenuated Total Reflection Infrared) Spectroscopy</u>

This technique is a non-destructive approach is based on the combination of IR spectroscopy with reflection theories, allowing the spectroscopic analysis of surfaces and is particularly efficient for the recognition of functional groups of natural and synthetic polymers.
The frequency range used is 4000-600 cm^{-1} and falls in the mid-infrared region).
The technique has numerous advantages, such as:

- Direct analysis of solids.
- High quality of the spectra for qualitative analysis.
- Small sample surface required for analysis.
- Non-destructive analysis.
- Short sample preparation times.
- Spectra independent of the sample thickness.
- High reproducibility.
- Wide variety of analysable samples (polymers, fibers, pastes, powders, suspensions, etc).

In terms of the use of the technique in the tanning sector, this is particularly efficient for the characterization of finishing polymers, for the recognition of some organic molecules, such as some plasticizers and fats, for the recognition of some anions and functional groups related to use of organic substances functionalized in chemicals, as well as for investigations on collagen and its possible modifications, following degradation phenomena.

Optical microscopy (OM)/ Phase-contrast microscopy (PCM)

Optical microscopy, which involves the use of light with a wavelength from near infrared to ultraviolet, covering the entire visible spectrum, is an indispensable approach for the observation and study of leather samples of different origins and intended use, thanks to the possibility of being able to investigate the morphological conformation, the distribution of the constituents, the fibrous microstructure, the chemical-physical characteristics; The optical reflected light microscope or stereomicroscope represents a useful tool for a first cognitive investigation and is a non-destructive approach, through which it is possible to obtain information on the morphology and on the three-dimensional surface structure; the phase contrast optical microscope exploits the phenomenon of light interference and uses a system capable of converting the differences in the refractive index into a difference in brightness; through this approach it is possible to obtain detailed information on the structural characteristics of the leather section.

Scanning Electron Microscope equipped with X-Ray Probe (SEM-X-Ray Probe)

The technique utilized is based on the electron-sample interactions and allows to reach information about the sample including external morphology and chemical composition. More in detail, The scanning electron microscope (SEM) uses a focused beam of high-energy electrons to generate a variety of signals at the surface of a given specimen. The signals deriving from electron-sample interactions reveal information about the sample. These signals include secondary electrons, backscattered electrons (BSE), photons (characteristic X-rays), visible light, and heat.
Secondary electrons and backscattered electrons are commonly used for imaging samples: secondary electrons are most valuable for showing morphology and topography on samples, while backscattered electrons are most valuable for illustrating contrasts in composition in multiphase samples, for example for rapid phase discrimination. X-rays are useful for microanalysis; their generation is produced by inelastic collisions of the incident electrons with electrons in discrete shells of the sample atoms.
Concerning leather defects investigations, the technique is potentially effective for both surface morphological/ topographic analysis and for microanalysis of the chemical composition of the surface.

Differential Scanning Calorimetry/ Thermogravimetric Analysis (DSC/TGA)

Thermal analysis allows to study the chemical-physical transformations undergone by the material during a heat treatment.
The "Thermogravimetric Analysis" (TGA) is a technique used to determine the mass variations of the sample in a controlled atmosphere as a function of temperature and time. Its usefulness is limited to transformations involving mass variations.
Scanning Calorimetry (DSC) is a thermal method in which the difference between the heat flows in the substance and in a reference is measured as a function of the sample temperature while both are subjected to a controlled temperature program.
In the tanning sector, the technique is particularly promising for investigations on the structure and technical features of leather and finishing polymers, allowing to analyse very small amounts of samples.

Gas chromatography-mass spectrometry (GC-MS) equipped with autosampler for VOC analysis

Chromatographic approaches are chemical-physical methods of separation of analytes mixtures based on the different affinity of each substance with two different phases, a stationary phase and a mobile phase, with which the mixture is put into contact. The sample is introduced into the mobile phase, wich will transport it inside the chromatographic system. By sliding the mobile phase over or stationary phase, competitive balances are established between the two phases and the substances that make up the mixture: therefore, the components of the sample enter the chromatographic system together but exit separately. When the mobile phase is a gas, the technique is called gas chromatography.
One of the most used gas chromatographic systems for the most transversal applications is Gas chromatography-mass spectrometry, used to study liquid, gaseous or solid samples; in thi approach, once the components leave the GC column, they are ionized and fragmented by the mass spectrometer using electron or chemical ionization sources; then ions are separated based on their different mass-to-charge (m/z) ratios. When a complex sample is separated by GC-MS, it will produce many different peaks in the gas chromatogram and each peak generates a unique mass spectrum used for compound identification. In that way, using extensive libraries of mass spectra, unknown compounds and target analytes can be identified and quantified.
In the tanning sector, many organic analytes potentially responsible for leather problems can be identified (antioxidants, plasticizers, hydrocarbons, etc.).
It is also possible to use a sampling system of the Purge and Trap type. It consists of the gas chromatographic analysis of the substances extracted from an inert gas, pre-concentrated in an adsorbent trap and cryogenically focused. This sampling system allows to determine volatile substances of particular interest in the tanning sector, which can provide useful information on the origin of specific defects.

Novel Non-Destructive Testing

Non-destructive testing (NDT) methods are defined as a set of technologies to verify processes and products in accordance with suitable quality control requirements. Any NDT method should preferably guarantee some important technical prerequisites, such as non-invasiveness and the ability to perform non-contact full-field measurements. There is a wide variety of NDTs used in very different sectors. However, NDTs relate more to industrial processes and products. Examples of applications include checking the integrity of materials and components, measuring geometrical properties, detecting and assessing damage and defects or even monitoring process parameters, where some novel approaches will be shown for applications in the tanning sector in monitoring defects and quality features of leathers.

2 Materials and methods

90 cases of defects were analysed: each case concerned leather and/or leather articles of different origin and for different intended uses (furniture, automotive, footwear, clothing, leather goods); more in detail: 30 cases for Furniture; 30 cases for Footwear; 10 cases for Leather Goods; 10 cases for Clothing; 10 cases for Automotive. Where deemed necessary, chemical products used in the production process were also analysed, as well as products and materials other than leather used to assemble the finished articles (adhesives, fabrics, synthetic linings).

The samples were analyzed with the following instruments:

- *Spectrum One ATR-IR (Attenuated Total Reflection Infrared)* Spectrometer (Frequency range: 4000-600 cm^{-1}).
- *Wild Heerbrugg* Stereoscope equipped with *OPTIKA* Room B3 Digital Camera.
- *Zeiss - Phoni 3* - Optcal Microscope (0.01 - 1 mm - Resolving power: 0,001 mm).
- *Zeiss EVO MA10* Scanning Electron Microscope (Magnification: 10x - 200,000x Resolving power: 2µm) equipped with *INCA X-act ENERGY-250XT* detector.
- Gas-chromatograph *TRACE 1300-THERMO SCIENTIFIC* coupled to a *TSQ 9000-THERMO SCIENTIFIC* mass spectrometer detector and equipped with *ATOM XYZ-TELEDYNE TEKMAR* autosampler for analysis of volatiles on solid and liquid samples.
- *STDQ600 TA* Instrument for DSC/TGA (Differential Scanning Calorimetry/Thermogravimetric Analysis equipped with a *THERMOSTARTM* Mass Spectrometer.
- *Electronic Speckle Pattern Interferometry System* (Laser 0.5W, wavelength 532nm) - *KA Associates, Coventry, Storrs, CT* (USA)

3 Results and Discussion

3.1 Main diagnostic techniques for defects detection

The 90 cases of defects were analysed using the main diagnostic approaches used to search for the causes of leather problems [2,3,4,5,6,7,8]. The cases investigated were chosen in such a way as to have a case history as representative as possible of all the main categories of defects found on the leather, as can be seen from table 1, showing an overview of the type of defects found and their incidence by category of article and their overall impact.

TYPE OF DEFECT		INCIDENCE OF THE DEFECTS BY TYPE OF LEATHER INTENDED USE [% incidence]					INCIDENCE BY TYPE OF DEFECT FOR 90 TOTAL CASES [% incidence]
		FURNITURE (30 CASES)	FOOTWEAR (30 CASES)	LEATHER GOODS (10 CASES)	CLOTHING (10 CASES)	AUTOMOTIVE (10 CASES)	
STAINS	STAINS from metals	23,3	3,3	0,0	0,0	10,0	10,0
	STAINS from organic substances	6,7	10,0	30,0	10,0	0,0	10,0
	STAINS from soaps	6,7	10,0	0,0	0,0	10,0	6,7
	STAINS from saline repousse	0,0	10,0	0,0	10,0	10,0	5,6
	STAINS from fatty repousse	0,0	6,7	30,0	40,0	0,0	10,0
	STAINS from microbial attacks	0,0	6,7	30,0	20,0	0,0	7,8
DIFFUSED CHROMATIC ALTERATIONS	Yellowing	26,7	0,0	0,0	0,0	20,0	11,1
	Other alterations	16,7	3,3	0,0	0,0	0,0	6,7
LACK OF DYEING HOMOGENEITY INSIDE THE LEATHER SECTION		6,7	3,3	0,0	0,0	10,0	4,4
DETACHMENT OF THE FINISHING		6,7	26,7	0,0	0,0	20,0	13,3
MORPHOLOGICAL DEFECTS AND CRACKS IN THE FINISHING		6,7	13,3	0,0	0,0	20,0	8,9
LOW MECHANICAL RESISTANCE		0,0	6,7	10,0	20,0	0,0	5,6

Tab. 1. Incidence of defects by analysed cases

The main techniques most commonly used to characterize defects and identify their causes, for each defect category, are shown in table 2. As shown in the table, and as a result of previous works, the resolution of individual cases required the simultaneous use of more than a diagnostic technique, even if some of these have proved to be decisive for characterizing specific critical issues.

TYPE OF DEFECT		% INCIDENCE BY TYPE OF DEFECT FOR 90 TOTAL CASES	MAIN DIAGNOSTIC TECHNIQUES
STAINS	STAINS from metals	10,0	SEM-X-Ray Probe
	STAINS from organic substances	10,0	GC-MS/ATR-IR
	STAINS from soaps	6,7	SEM-X-Ray Probe/GC-MS
	STAINS from saline repousse	5,6	SEM-X-Ray Probe/ATR-IR
	STAINS from fatty repousse	10,0	GC-MS/ATR-IR
	STAINS from microbial attacks	7,8	SEM/OM/GC-MS
DIFFUSED CHROMATIC ALTERATIONS	Yellowing	11,1	ATR-IR/GC-MS/SEM-X-Ray Probe
	Other alterations	6,7	GC-MS/SEM-X-Ray Probe/ATR-IR
LACK OF DYEING HOMOGENEITY INSIDE THE LEATHER SECTION		4,4	SEM-X-Ray Probe/PCM
DETACHMENT OF THE FINISHING		13,3	ATR-IR/GC-MS/OM/PCM
MORPHOLOGICAL DEFECTS AND CRACKS IN THE FINISHING		8,9	PCM/OM/DSC-TGA/ATR-IR
LOW MECHANICAL RESISTANCE		5,6	PCM/OM/DSC-TGA

Tab. 2. Main techniques vs type of defects

The main contribution of each of these techniques in the characterization of defects and in the identification of their causes will be discussed in the following**.**

3.1.1 The role of ATR-IR (Attenuated Total Reflection Infrared) Spectroscopy

As found from the examination of the cases covered by this study and from previous works [2,3,4,5,8,13], the technique was mainly used to solve problems related to finishing, in relation to critical issues related to mechanical, chemical and chemical-physical finishing fragility. Examples are represented by cases in which the fragility towards some chemical agents, deriving from the leather or from the assembly materials migrated in the finishing, has determined the damage of some articles, due to the chemical nature of the finishing. For example, figure 1 shows the overlap of spectroscopic profiles of samples that have exhibited a different behaviour towards these agents: the profile of the damaged sample is specifically found to be characteristic of nitrocellulose-based finishes, more sensitive to the action of organic-polar solvents, compared to finishes based on polyurethanes.
More in detail, Figure 1, shows that the blue profile is typical of polyurethanes, given the presence of the following typical signals**:**

- Stretching vibration -NH at 3300 cm^{-1}
- Stretching vibration -C=O a 1720-1726 cm^{-1} (amide I)
- Bending vibration -NH at 1530 cm^{-1} (amide II)

The green profile is typical of nitrocelluloses, given the presence of the following typical signals:

- Two strong signals of symmetrical and asymmetrical stretching vibrations of -NO_2, at 1640-1650 cm^{-1} and 1275-1278 cm^{-1}
- Stretching vibration of -C–O, of glucopyranose ring at 1070 cm^{-1}
- Stretching vibration of -O–NO_2, at 840 cm^{-1}

Other cases have shown that both finishes consisting of nitrocellulose have shown a different behaviour to the action of protic solvents, due to the different nitration degree of nitrocelluloses, identified by the different intensity of the signals relating to the -NO_2 groups.

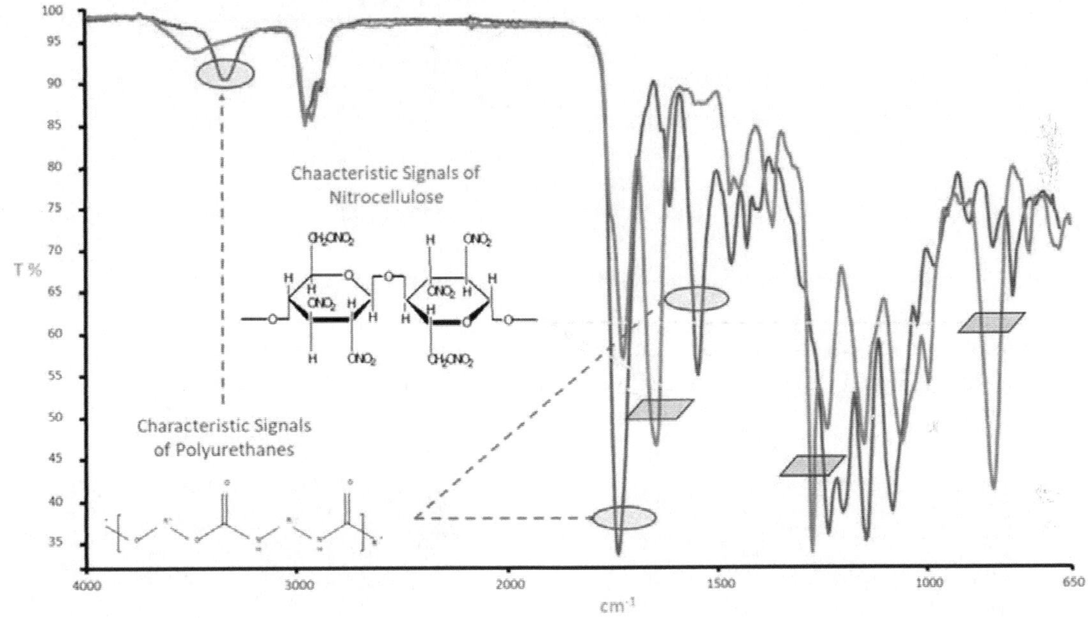

Fig. 1. IR spectra overlap of the surface of finished leather samples with two different polymers

The technique has also been effectively used to study cases of defects related to surface yellowing and related to the use of aromatic polyurethanes in finishing (Fig. 2), where the tendency to yellowing of these categories of polymers under given conditions of thermo-climatic aging (exposure to light and heat) is known[8,11,12].

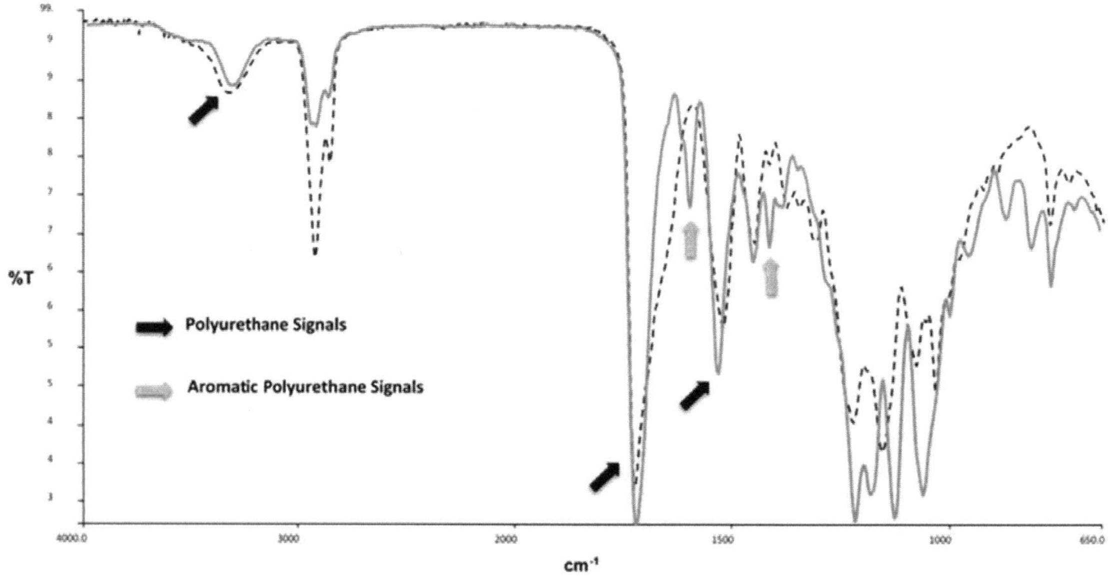

Fig. 2. IR spectra overlap of the topcoat of defective (yellow) and not defective (black) samples.

3.1.2 The role of Optical microscopy (OM)/ Phase-contrast microscopy (PCM)

Stereomicroscopy was proved to be generally effective for all preliminary surface analysis, in order to have some initial clues on the possible origin of defects, mainly concerning grain and finishing damage (Fig. 3).

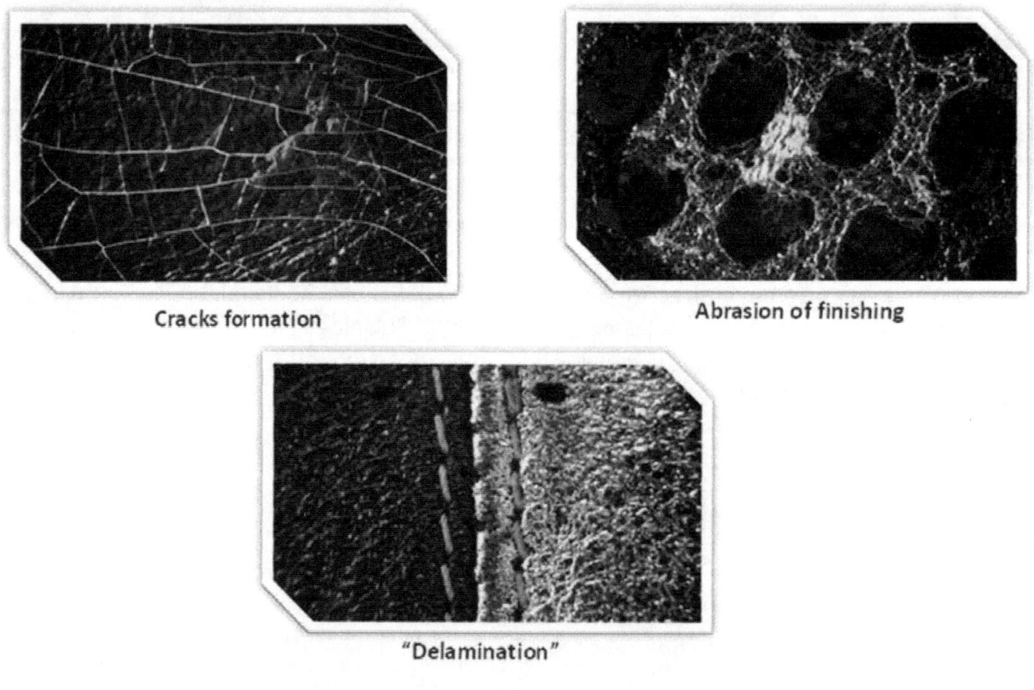

Fig. 3. Images obtained from the stereomicroscope of the surface of defective samples

The use of Phase-contrast microscopy (PCM) for investigating the possible causes of mechanical-structural finishing defects, through the examination of the leather section was then proved to be very decisive (Fig. 4).

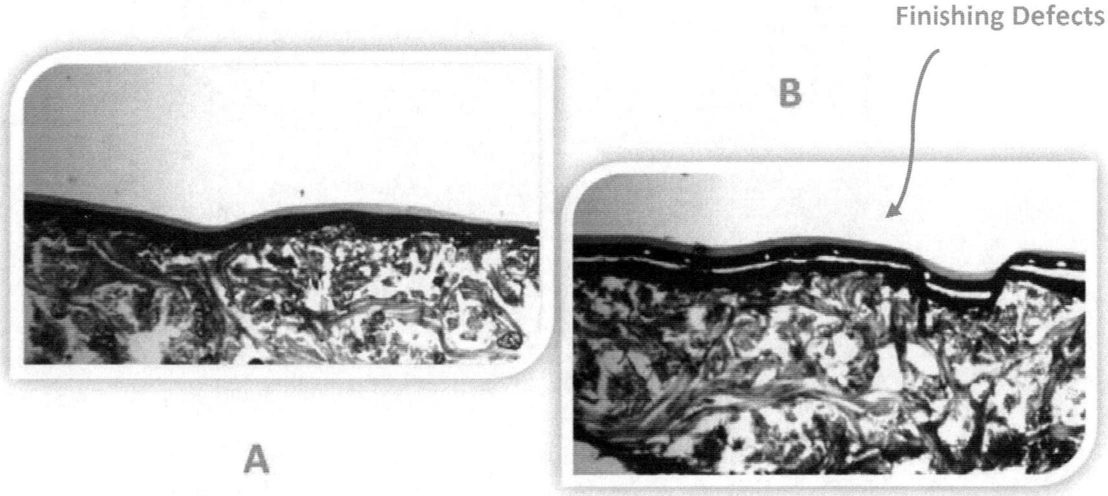

Fig. 4. Images obtained by the phase contrast microscope,
of the section of two samples of patent leather (not deteriorated - A; deteriorated - B)

This technique can also provide the first clues on possible causes of defects related to process errors, allowing to identify the presence in section of hair residues, blood residues and leaks in the structural characteristics of the leather fibres.

3.1.3 The role of Scanning Electron Microscope equipped with X-Ray Probe (SEM-X-Ray Probe)

This technique has proved to be one of the most useful and transversal for the study of leather defects, almost always proving crucial for the understanding of the possible causes and phenomena underlying the origin of the critical Issues[6].
Concerning the morphological and topographical analysis, the technique allowed the investigation of the crystalline, rather than amorphous morphological features of deposits, suggesting their possible nature (saline, fatty, organic, etc.) (Fig. 5).
Beside the study of defects, morphological analysis allowed the characterization of animal species, when it has not been possible by optical microscopy, that was in all those cases in which the leather surface was heavily damaged or in slightly grounded leathers, where grain design was hardly recognizable.
Finally, the topographic analysis of leather surface was proved to be suitable for finished leathers affected by peculiar defects, due to the presence of bumps, holes, bubbles folds and halos.

Fig. 5. SEM morphological analysis of leather defects

SEM analysis was also fundamental for the study of defects caused by microorganisms: there are in fact many microorganisms that can cause damage to the leather; the typology and the amount of microorganisms found, resulted to vary according to some environmental factors, as pH, temperature, water activity, concentration of biocides and so on. As found also in previous studies (ref), microorganisms found on leather belong both to *Prokaryota* (as *bacteria*) and to *Eukaryota* domain (as *moulds* and *yeasts*, from *Fungi* kingdom). However, the most commonly encountered microorganisms was proved to be wide set of *fungi* were found, with particular reference to *moulds*. They are recognizable through peculiar spores, as the moulds belonging to *Ascomycetes* class, with particular reference to the ones from genus *Penicillium* and *Aspergillus*. Most of these can be found in pickled, wet-blue and finished leather, resulting to be resistant at high temperatures and to have a quick growth in presence of an high humidity degree. These *moulds* differ by colour as much as by appearance; normally ones from genus *Penicillium* are glassy or velvet looking and are green or blue coloured, while the ones from genus *Aspergillus* form yellow or black stains, where SEM analysis resulted to be effective for the recognition of both genus (Fig. 6). Most of stains produced by moulds formations on leather among all the tanning process resulted to be blue, gray or black coloured. The blue formations have been normally identified as *Penicillium glaucum*, the gray ones as *Mucor mucedo*, while the black ones have been classified as *Aspergillus niger*.

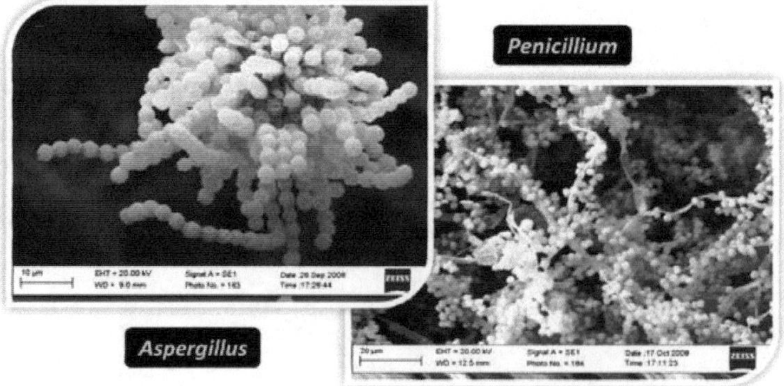

Fig. 6. SEM study of defects caused by microorganisms

When other techniques are coupled to the SEM analysis, such as GC-MS, it is also possible to get clues to the state of degradation of the leather, which suggests the exposure of the articles to climatic conditions favourable to the development of microorganisms, and it is also possible to obtain information on the biocides used in the process, whose quantity and type, can play a certain role in fostering microbiological attacks in leather articles.

Finally, SEM-Microanalysis, was found to be one of the most effective and crucial technique, particularly to study the possible origin and causes of metal-based defects; during the tanning process, there are many metals that can come into contact with the leather, both accidentally and due to their use often foreseen in the processing, where they can be involved in leather defects, when able to determine:

- coloured spots on the leather surface, of different size and appearance, sometimes in combination with deposits;
- non-homogeneity in the dye distribution, both on the surface and across the leather section.

These main types of defects are generally due to formation mechanisms which can be:

- direct type, through the formation of coloured metal compounds (for example oxides, sulphides, etc., of metals such as lead, copper and iron or coloured compounds deriving from interactions of some metals such as iron and magnesium with vegetable tanning agents, as well as from more complex interactions between metal and organic substances, which can occur at the finishing level);
- indirect type, in the event that metals (commonly chromium and calcium) are involved in the formation of compounds that can produce a non-homogeneous absorption of tannery chemical products, with particular reference to the fattening and dyeing phases. In some cases, these mechanisms can be complex and difficult to identify.

Figure 7 shows some metal-based defects investigated by SEM-Microanalysis.

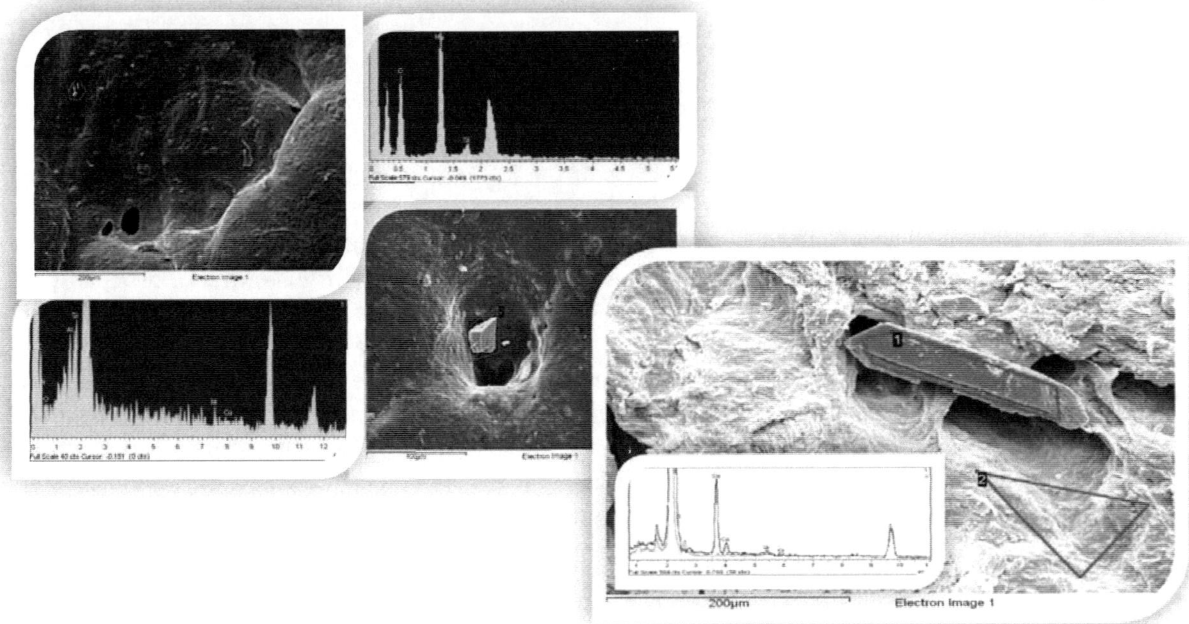

Fig. 7. SEM-Microanalysis of metal based-defects

3.1.4 The role of Differential Scanning Calorimetry/ Thermogravimetric Analysis (DSC/TGA)

This technique was proved to be effective both for studying defects related to possible critical issues of finishing polymers, and defects related to structural issues of the components of the leather, such as collagen itself, when stabilized in a non-effective way.

With reference to the defects of finishing polymers, the knowledge of the degree of crystallinity of the polymers, the determination of the glass transition temperature, and more generally of the parameters correlated to their chemical, structural and consequently performance characteristics, has represented a crucial tool for the comparative study of finishes and for the design of improved and innovative solutions (Fig. 8).

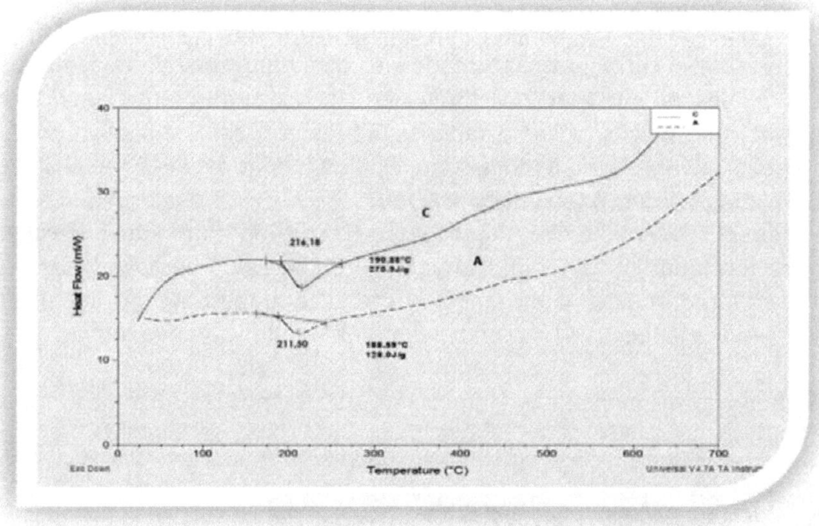

Fig. 8. DSC study of two finishing samples based on nitrocellulose characterized by a different degree of nitration

These diagnostic approaches have also proved to be potentially strategic for the characterization of possible defects related to the thermal resistance of new generations of leathers, particularly in relation to the possibility of carrying out comparative studies of thermal stability of leather processed with different types of tanning agents; these studies complete and enrich the macroscopically obtained information by determining the shrinkage temperature of the leathers (according to UNI EN ISO 3380: 2015); more in detail, differential scanning calorimetry (DSC) allows to study the enthalpy changes associated with the denaturation of collagen, of which contraction represents the macroscopic manifestation. These changes are associated with other chemical and mechanical characteristics of the material, as a result of the breaking of non-covalent bonds that stabilize the collagen with the increase in temperature.

3.1.5 The role of Gas chromatography-mass spectrometry (GC-MS) equipped with autosampler for VOC analysis

The use of Gas chromatography-mass spectrometry (GC-MS) was proved to be particularly effective in all those cases of defects caused by organic substances; more in detail, many stains or diffused chromatic alterations resulted to be related to the migration of organic pigments or fatty substances, directly involved in the formation of defects; furthermore, other organic substances, such as antioxidants, plasticizers, fats, biocides, whose presence was revealed with this technique, were involved in the defects, through more indirect and more complex mechanisms.

Moreover, when combined with autosampler for VOC analysis, this technique allowed to reach also Very crucial information for the resolution of some cases of defects; in more detail, the analysis of VOCs made it possible to identify two main classes of compounds capable of providing clues to the origins of specific critical issues:

- aldehydes, ketones, furans, sulphur compounds, which can provide indications on the state of degradation of the organic components of the leather, such as fats, in order to clarify the possible contribution of thermo-climatic aging conditions on the onset of defects related to degradative phenomena of leather, finishing or involved in the development of microorganisms;

- solvents deriving from leather, such as production residues or deriving from assembly materials, in order to identify the causes of degradation of some components of the finishing sensitive to the action of these solvents (Fig. 9).

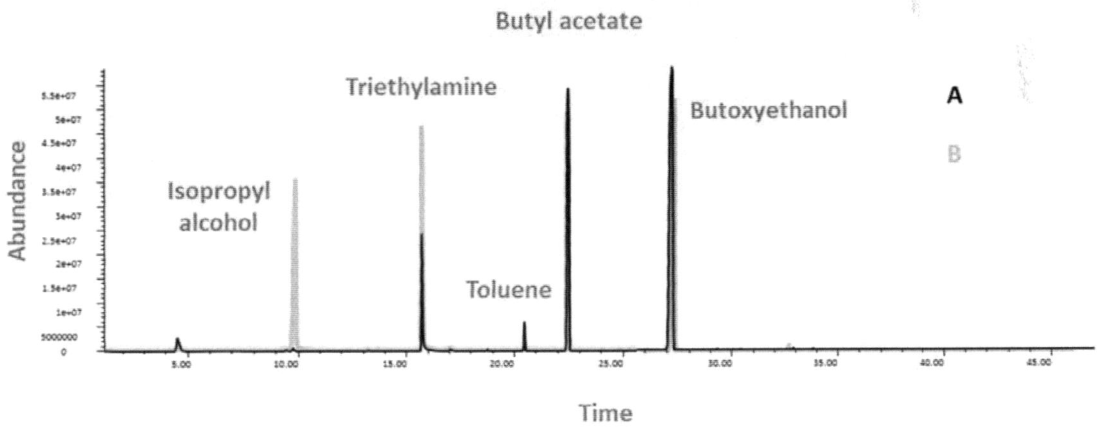

Fig. 9. VOC comparative analysis for the evaluation of residual solvents in two different samples

3.2 Main Novel Non-Destructive Testing for the detection of leather defects

Different NDT methods have been developed to meet different needs [14], some of which have proved suitable for the timely analysis of leather defects. Among many others, optical interferometric techniques — using lasers — have found use in the testing of various types of composite materials and textiles. This type of technology, thanks above all to the evolution of sensors, the improvement of calculation capacities and the increased efficiency and compactness of laser sources, offers the possibility of real-time control and monitoring in full field on articulated surfaces such as those made of leather objects. Interferometric techniques are based on the interference phenomenon and, in particular, on the so-called 'speckle'. Speckle — which gives its name to one of the most widely used techniques [15], Electronic Speckle Pattern Interferometry (ESPI) — is a phenomenon based on the coherence of the light used (i.e. the laser), which when striking a rough surface, makes a random phase jump without losing its coherence. This generates an interference effect that can be constructive or destructive, resulting in images composed of dark and bright 'dots'. This dot pattern is used to detect surface variations with a sensitivity below the wavelength of the laser used (i.e. in the visible range of a few hundred nm). The basic scheme of the optical setup is shown in figure 10.

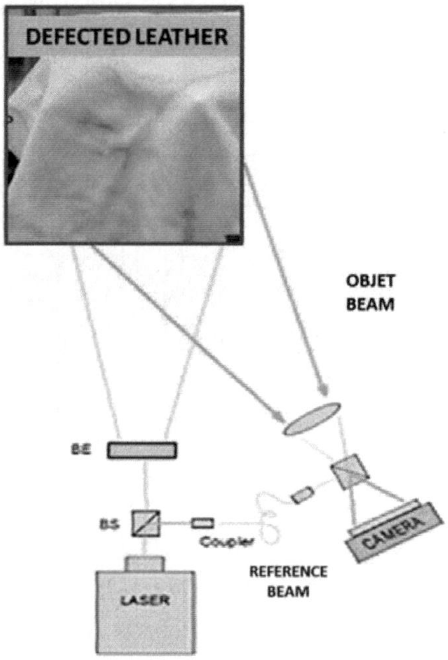

Fig. 10. Basic scheme of the optical setup of a speckle system

If the object to be examined is subjected to adverse conditions, e.g. through heating, it changes its geometric properties and consequently the way it reflects light, i.e. its speckle. The digital comparison of the different speckles leads to the creation of an image showing a series of overlapped fringes on the screen of the surface being examined. By observing the shape and direction of the fringes, it is possible to highlight surface inconsistencies caused by damage such as cuts or incisions, leather discolouration, scratches or curling. An example of an evaluation of a piece of leather (right) is shown in the figure 11, where from the processed image of the fringes (left) it is possible to highlight the presence of barely visible scratches and cuts on the surface of the leather.

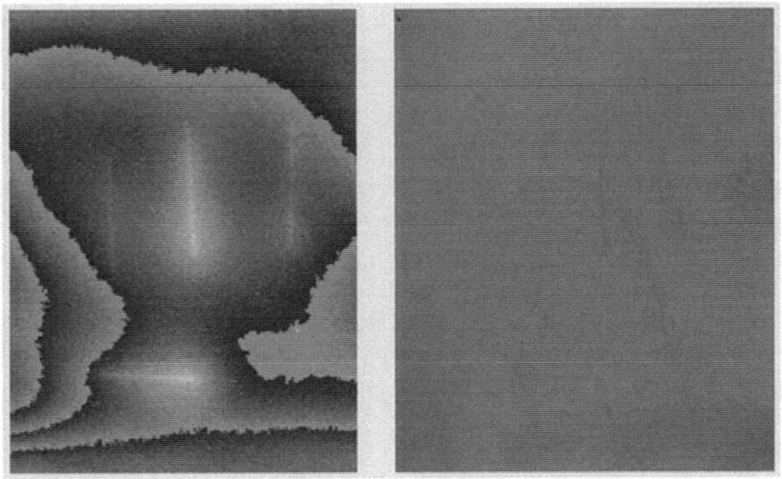

Fig. 11. NDT of a slightly defective leather sample

In addition to techniques based on interferometry, there are also computer vision techniques, which are based on the scanning and processing signals and images at different wavelengths, from the visible to the infrared (NIR, SWIR etc.). More generally, a computer vision system is a set of hardware and software used to equip a machine with functions that are typical of human vision. The hardware used for two-dimensional imaging involves the use of cameras, based on CCD or CMOS sensors, which can be a) linear, scanning a single line at a time but also able to scan a large number of lines per second and with photosensitive elements that tend to be large, but which require an auxiliary external movement and trigger system to form a 2D image; b) frame, directly scanning a two-dimensional (2D) image and operating at slower speeds than linear cameras and with pixel dimensions that can also be particularly small. Cameras can scan a variable number of channels: one for monochrome cameras, three for colour cameras (Red, Green, Blue), many more for multi-spectral or hyperspectral cameras. Depending on their sensitivity to different wavelengths, the channels may cover the visible spectrum, the infrared (near or far) or wide frequency ranges as in hyperspectral techniques, which associate each of the many channels with a specific, very narrow band of wavelengths. The choice of sensor type depends on the type of behaviour or defect you wish to detect and the conditions under which the images can be captured. In the tanning industry, but also in all industrial applications, it is necessary to make the data acquisition and processing operations in line, so robotic systems specifically designed for handling, scanning and analysing products in line are often used. These systems include automated guided vehicles (AGVs), which can operate autonomously to move sensors and N-degree-of-freedom robotic arms, i.e. they can be used to handle products and to position sensors so that even very large areas can be inspected.

Finally, at the software level, a vision system must be able to analyse the acquired data in order to extract relevant information, in this case, the presence and possible characterisation of a defect. Historically, image analysis has been carried out by defining, extracting and identifying particular distinguishing features of a product in order to analyse colour configurations, surface structure or otherwise homogeneous and non-homogeneous regions of the image. Once the features have been extracted, a classification algorithm can recognise which areas of the image are defective or not, based on the numerical values of the extracted features. In this context, in recent years, the advent of deep learning techniques [10,16], a family of artificial intelligence models that are particularly effective for image analysis, has made it possible to tackle classification problems (and therefore also NDTs) with a new approach based on artificial and data-driven neural networks, thanks to the ability of these models to learn from the data autonomously which are the best features to extract. These models are generally trained in a supervised manner on so-called labelled datasets, i.e. with images for which the defect or no defect classification is known and available a priori. Under appropriate conditions, these algorithms are able to generalise what they have learnt and thus recognise and characterise defects even when required to process new (in-line) images that were never seen in the training phase. This feature is particularly important in applications, such as NDT of leather, in which the intrinsic variability of the material, of natural origin, makes constructing an exhaustive case history of the types of defects and their possible manifestations extremely complex. In this case, having algorithms capable of autonomously constructing sufficiently general representations of both the defect-free material and the different classes of anomalies constitutes a powerful tool for solving NDT problems.

4 Conclusion

The use of advanced diagnostic techniques considered, mainly non-destructive or based on the use of small amounts of samples, allows to obtain crucial information for understanding the phenomena capable of originating the most disparate leather defects; specifically, the use, also conjugated of spectroscopic methods, of specific gas-chromatographic techniques, of optical and electronic microscopy approaches and of systems for thermal analysis, allows to identify errors and critical issues deriving from the specific phases of the production process, in order to plane improvement actions and designing more efficient and innovative solutions to guarantee product quality.

Moreover, it is possible to act in a preventive and predictive manner through the increasing use of NDT technologies and automated approaches for the in line control of production: interferometric optical techniques, and more generally computer vision techniques, are suitable for fast inspections and were found to be consistent with future trends in NDT technologies, which will see the development of automated non-destructive testing in the coming years thanks to the increasingly widespread use of artificial intelligence (deep learning) and robots. The tanning industry and the production chain can certainly benefit from these technological developments in order to achieve ever higher standards of quality and functionality of products that use leather.

5 References

1. *Branca, Antonella, et al. "Automated system for detection and classification of leather defects." Optical Engineering 35.12 (1996): 3485-3494.*
2. *Florio C., Calvanese G., Naviglio B. - The dark side of leather. Finishing issues and light-dependent properties of pigments - XXXIII IULTCS CONGRESS, Novo Hamburgo, (BRASILE), 24-27 Nov. 2015, PDF 191; CPMC, 2016, 92(2/3), 87-94.*
3. *Florio C., Calvanese G., Naviglio B., Mascolo R., Grasso G. - Finishing defects. A special case concerning pigmented finishing based on TiO_2 - CPMC, 2009, 85(3), 131-143.*
4. *Florio C., Mascolo R., Calvanese G. - Patent and laminated leather: the role of crucial parameters in the damage of fashion goods - XXX IULTCS CONGRESS, 365, BEIJING (CINA), 11-14 ottobre 2009; ARS TANNERY, 2(3/4), 57-61/62-67(2011).*
5. *Florio C.; Mascolo R.; Calvanese G.; Naviglio B. - What causes laminated leather defects - Leather International, 2010, 212 (4799): 24-28 - Tecnica Calzaturiera, November 2010: 42-44.*
6. *Florio C., Naviglio B., Calvanese G. - Advanced Approaches in Leather Research using X-Ray Probe Equipped SEM Microscopy - JSLTC - Volume 101 - July-August 2017 - Number 4 - ISSN 0144-0322, 165-172.*
7. *Florio C., Calvanese G., Naviglio B., Siani B., Metal based defects - Resolving actions and their planning through the instrumental diagnosis of the semi-processed leathers - Cuoio Pelli e Materie Concianti (CPMC), 88, 5, 261-276, 2012; LX AICC National Congress, Ravello (SA) - June 15 2012.*
8. *Florio C., Aveta R., Calvanese G., Naviglio B. - Advanced diagnostics and innovative solutions for leather defects: the problem of yellowing. JSLTC - Journal- Society of Leather Technologists and Chemists - December 2019. 103(6):296-304.*
9. *Kwak C., Ventura J.A., Tofang-Sazi K. - Automated defect inspection and classification of leather fabric, November 2001Intelligent Data Analysis 5(4):355-370.*
10. *Liong, Sze-Teng, et al. "Leather defect classification and segmentation using deep learning architecture." International Journal of Computer Integrated Manufacturing 33.10-11 (2020): 1105-1117.*
11. *Michel A., The many causes of yellowing, Leather International, 207(4756-april), 64(2005).*
12. *Michel P.B., Yellowing on footwear and leather goods, CTC Entreprise, 21(10), 6-9(2005).*
13. *Naviglio B., Calvanese G., Florio C. - The characterization of chemicals used in finishing - XXIX IULTCS Congress, Washington (USA), 21-24, June, 2007 - CPMC, 2007, 83 (4): 225-245.*
14. *Pagliarulo V., Ferraro P. et al. "Impact damage investigation on composite laminates: comparison among different NDT methods and numerical simulation" Meas. Sci. Technol. 26, 085603 (9pp) (2015).*
15. *Pagliarulo, V., Ferraro P., et al. "Combining ESPI with laser scanning for 3D characterization of racing tyres sections" Opt. Lasers Eng. 104 (2018), pp. 71-77, 10.1016/j.optlaseng.2017.07.004.*
16. *Pazzaglia Giulia, et al. "A Deep Learning-Based Approach for Automatic Leather Classification in Industry 4.0." Pattern Recognition. ICPR International Workshops and Challenges: Virtual Event, January 10–15, 2021, Proceedings, Part IV. Springer International Publishing, 2021.*

XXXVI IULTCS CONGRESS

3rd – 5th November 2021, Addis Ababa, Ethiopia

CARBON FOOTPRINT ITALY: A COMMUNICATION TOOL SUPPORTING ENVIRONMENTAL AND CLIMATE EFFORTS

Daniele Pernigotti[1]

Aequilibria Srl – SB

[1] Corresponding author addresses:
Phone: +39(0)418653238
E-mail: dpernigotti@aequilibria.com; info@aequilibria.com

Abstract

This work reviews the opportunities to make the communication of Carbon Footprint of Products (CFP) more effective, starting from the experiences of Carbon Footprint Italy (CFI) and Carbon Footprint international.

In view of the implementation of the ambitious policies set globally, requiring a quick and irreversible decarbonization of all the sectors, it is crucial to guarantee that transparent, reliable, and effective information is provided to the public.

The ISO 14067 standard is the internationally recognized tool for the CFP quantification. Nowadays, it is broadly adopted by companies aiming at providing reliable results.

The analysis is focused on CFI, the Italian initiative. First, it values credibility, only allowing the participation to products/organizations owning a verification statement issued by an accredited independent third-party. It promotes transparency, making all the information publicly available on a dedicated webpage for each registered product, easily accessible through a QR code that can be printed on the product itself.

In addition, it valorizes reliability by using the blockchain technology, that stores the data in an unmodifiable way, allowing for solid and credible performances tracking over time.

The review also covers the global scenario of existing initiatives, gathered in the international network created to boost the global recognition of carbon footprint.

Another aspect the analysis assesses are the opportunities for collaboration between different actors of the value chain, as crucial to encourage virtuous cycles. The example of the tannery and leather supply chain is reported, with involved realities ranging from raw materials suppliers, tanneries, and final product users.

Keywords: Carbon Footprint, climate, communication, sustainability, neutrality, leather, supply chain

List of Abbreviations

CFI	Carbon Footprint Italy
CFO	Carbon Footprint of Organizations
CFP	Carbon Footprint of Products
GHG	Greenhouse Gases
PO	Programme Operator

1. Introduction

Climate change is one of the most complex challenges that humankind has ever had to face. It is now evident how this challenge is becoming more and more compelling, and how the efforts must be global, quick, and resolute. This is confirmed by the recent publication of the contribution of Working Group I to the Sixth Assessment Report of the United Nations Intergovernmental Panel on Climate Change, namely the *Climate Change 2021: The Physical Science Basis,* that thoroughly analyses the physical basis of the climate systems, provides evidence of how climate is changing and how this is strictly related with human activities.

The awareness of this urgency is not only widespread within the scientific community, but within all the sectors of our society. More and more ambitious policies are being adopted all over the world, regulating climate actions, setting ambitious targets, and boosting virtuous actions and market mechanisms. A remarkable example is represented by the European Green Deal, a comprehensive policy that sets carbon neutrality goals, and recognizes the relevance of integrated actions from several sectors (as climate, industry, energy, transport, agriculture, etc.).

This revolution is also occurring at a social and economic level. In this context, the market urges companies to embark serious actions towards sustainability. Massive investments to reduce the impact on climate change are made. Many companies have already undertaken pathways in this direction, by using innovative efficient technologies and by adopting advanced environmental management systems. Nowadays, these efforts towards sustainability are becoming even more intense, and companies are introducing specific tools for carbon management within their strategies.

Several voluntary methodologies for the quantification of greenhouse gas emissions are already available on the market. Some of them are extremely reliable and internationally recognized, as for example the quantification of Carbon Footprint of Products (CFP) in accordance with the ISO 14067, or the Greenhouse Gas (GHG) inventory of organizations in accordance with the ISO 14064-1.

Anyway, implementing technical improvements and adopting carbon management strategies in accordance with the recognized standards is not enough to ensure the massive market adoption of products with a low carbon footprint; another key element is the effective communication of the results achieved, combined with a close relationship between producers and consumers.

It should be noticed that, without the presence of a recognized Programme[2] dedicated to the communication of carbon footprint, it is not easy for companies to select the proper pathway to be followed to ensure a serious communication of the efforts undertaken. In fact, they can opt for using self-declarations, that are not easy to justify in a credible way to the market, even if based on trustworthy efforts. They thus risk to fall into the so-called *greenwashing*, either consciously or not. At the same time, the same difficulties are faced by consumers or clients, to understand how serious the information communicated is, and thus to take informed decisions.

Hence, there is a need for clarity in this sense, by using a serious, reliable, transparent but simple and effective CFP Programme able to complement the technical and management efforts, from a communication point of view (Vizzoto et al, 2021).

1.1 Objectives of the Study

In this context, in the past few years several Programme Operators (POs) on carbon management were launched, aiming at simplifying and regulating the complex communication framework characterizing the carbon management claims. These initiatives, nationally managed, started a collaboration pathway within the *Carbon Footprint International* network.

This work is focused on the Italian experience, *Carbon Footprint Italy*: it provides an overview of the innovative features that led it to be the first initiative ensuring such level of guarantees and reliability on the market, due to its flexibility and its ability to follow the market needs, potentially involving all the actors

[2] Also known as Programme Operator, or PO.

along the leather value chain. In addition, it is also taking advantage from the introduction of best available technologies that provide a high-quality tracking of the reduction over time of the carbon emissions in a guaranteed, reliable, and effective way, giving a new shape to the carbon management communication scenario.

1.2 Outline of the work

After an introduction on the main elements ensuring credibility and reliability to the CFI Programme, and of the innovative technologies implemented, this work is divided in four main parts:

- Overview of the communication context, often based on self-declaration where the risk of greenwashing is significant, and of the need for Programmes regulating and ensuring the reliability of the climate claims.
- Outline of the structure of Carbon Footprint Italy, including the rules on which it is based and the requirements to be fulfilled to register, the guarantees that it offers in the voluntary market and of the different types of labels available.
- Overview of the opportunities offered by CFI through the leather sector case.
- Presentation of the international network in which it operates, that is moving towards the mutual recognition of the single initiatives to promote the consistent implementation of these concepts globally.

2. Materials and Methods

Carbon Footprint Italy (CFI) is the Italian Programme Operator on Carbon Management. The goal is to enable companies to communicate credible and reliable information on their GHG inventory and on the Carbon Footprint of Products (CFP) of their products to all the actors of the value chain, and to the users and customers.

To participate in the Programme, companies must fulfill a set of requirements.

First, organizations and products shall own a verification statement issued by an accredited independent third party. The international standards that should be fulfilled to achieve the registration in CFI are the following:

- ISO 14067 for products (ISO, 2018);
- ISO 14064-1 for organizations (ISO, 2018);

- PAS 2060, for carbon neutrality statements (BSI, 2010).

The third-party verification represents a guarantee on the reliability of the values of CFP and of GHG inventory. Anyway, additional elements are needed to ensure an effective and transparent communication.

As mentioned, the first thing to be provided is a complete information. This is to be ensured considering that all the background information supporting the study cannot be directly depicted on the single marks that are printed on the products, because of space constraints and of clarity reasons.

That is why in CFI the use of a label is combined to a unique QR code. Each QR code is associated to a single registration realized in the Programme and is linked to the dedicated webpage on the CFI website especially created for the specific product or organization.

The webpage reports all the useful background information, that are relevant to be communicated, as indicated in the ISO 14026 (ISO, 2017). For example, for the Carbon Footprint of Products, the following data are reported:

- Organization name;
- Organization address;
- Phone number, website and contact person;
- Product name;
- Product description;
- Registration date;
- Declared/functional unit;
- Value of CFP. It is reported both as a comprehensive value, and as values split per type of emission, namely net fossil GHG emissions, biogenic GHG emissions, biogenic GHG removals, GHG from dLUC. In addition, the aircraft GHG emissions are reported also separately, as requested by the reference ISO standard;
- Reference year;
- System boundaries;

- Stages excluded from the lifecycle;
- Production plants included;
- CFP Study report name and version;
- Additional notes;
- Verification body;
- Link to blockchain;
- Copy of the verification statement.

Another technology adopted by CFI is blockchain. Blockchain technology enables actors to verify and audit transactions and track them in ledgers (Saberi et al, 2018). Once the transactions are approved, they can be added as a new block in the blockchain. Smart contracts then increase the credibility of transactions. In this way, values are assigned to the original information, and the system tracks them over time. It should be noted that, although new blocks can be added, it is not possible to delete the existing ones, that remain saved and tracked in the system. Blockchain technologies are thus becoming increasingly helpful to implement climate action.

In this way, thanks to this technology companies are made accountable for their CF claims through unmodifiable information reported in the Programme; consequently, it is also possible to control in a reliable way the evolution over time of the companies' GHG reduction pathways.

3. Results

The communication of climate data is becoming more and more important at the market level.

Many companies are embarking decarbonization pathways, by investing into innovative technologies, materials, engineering to develop low carbon products. It is crucial for them to combine the technical advancements with a proper communication of the commitments taken and of the results obtained.

A significant interest was in fact demonstrated by the market (Whelan et al, 2019), where the choices of consumers and clients are also depending on the level of engagement and credibility of the company offering a product. Because of this, in parallel, also the number of initiatives and labels available on the market has been quickly increasing.

This is creating a complex and rich framework, offering several possibilities and pathways to organizations willing to communicate their efforts. Within this broad availability there are also some companies that decide to go for self-declarations, that often are not related to recognized schemes, thus intensifying the confusion on these topics. The consequence is that there is a high risk of greenwashing.

The topic of voluntary disclosure of the commitments made and of the results obtained is thus becoming crucial, and so is the need of responding to these needs.

This cannot be addressed through self-declarations, but through a communication able to give guarantees to the markets and recognized by accredited third parties, namely the Programme Operators (POs). This is also indicated in the ISO 14026 on environmental labels and declarations, that provides requirements and guidelines for footprint communication Programmes.

For this number of reasons, CFI represents an important tool to guarantee that the communication of the efforts undertaken be clear, transparent, and certified. Carbon Footprint Italy also represents a credible and effective network, capable of pushing different actors, along the whole supply chain, to follow similar paths. It is, in fact, a powerful tool that involves and influences the decisions taken at different stages of the value chain.

3.1 Specific opportunities offered by the Italian PO

As mentioned above, this initiative represents one of the most interesting and innovative responses to the need for reliable communication tools highlighted by the market, applying technologies with the perspective of guaranteeing transparency and credibility (Evercity, 2020).

In detail, CFI aims at working with the broad definition of carbon management, that is not a univocal concept. In fact, several pathways and strategies are available and can be selected and combined, depending on the possibilities, on the specific sector, on the level of ambition, and thus on the stage of the decarbonization pathway in which a company is.

CFI was thus structured to reflect this complexity, offering the opportunity to track the actions undertaken and the results achieved at different levels.

At the beginning, the focus was on the quantification of carbon footprint. Two labels were available:

- CFP (in accordance with ISO 14067), for products;
- CFO (in accordance with ISO 14064-1), for inventories at organization level.

The interest of the market is quickly evolving, and it was noticed that quantification is now only the first step of a broader pathway. Many organizations are now committing to reduce their CFP/CFO. Therefore, a new label was introduced, for both products and organizations: carbon reduction. This allows to keep the enhancements monitored over time, also thanks to the blockchain technology, ensuring that serious efforts have been undertaken and that the results are noticeable. Another application is to monitor the GHG inventory over time, that can be also used to value the data disclosed with other international initiatives (as the Science Based Target initiative, or SBTi), that require organizations to yearly disclose their results.

Currently, more and more companies are committing to carbon neutrality: this led CFI to introduce an additional label, namely the carbon neutrality one, to give reliable information about these commitments, that is issued when the organization decides to offset the residual GHG emissions remaining after the mitigation actions.

4. Discussion

4.1 The leather value chain

The leather value chain is an excellent example to understand CFI potentialities, since several companies have adhered to the programme. They often represent at the same time customers and suppliers that, interacting among themselves, affect each other's market decisions.

Within it, CFI encloses:

- **Chemical pre-processing**: companies that produce chemicals used in tanneries for the processing of leather.
- **Leather manufacturing**: companies that process hides have developed not only CFP and EPD studies, but also drafted the PCR for "Finished bovine leather", which now defines the rules to be followed for all studies in the sector.

- **Use of leather**: companies that buy processed leather and use it to produce high-quality products, by focusing on the transparency of their suppliers. Some EPDs have been published, which are based on certified products registered on CFI.

As a matter of fact, it is unequivocal how customers are more and more asking for information of Carbon Footprints. Carbon Footprint Italy represents, in this market, the common point between the actors that recognize the reliability and professionalism of the studies made to take part in the PO. By doing so, they are pushed to reach the same level of seriousness and are encouraged to take the same measures. Companies already registered on CFI, on their side, will be led to choose to work only with actors that are sensitive to the same environmental issues, avoiding those who are not pursuing a decarbonization path. There is in fact a growing interest in moving towards Carbon Neutrality, especially in the leather sector. Suppliers that can seize the opportunity will stand out from competitors and gain an advantageous position on the market.

It thus appears clear how these mechanisms are generating a virtuous cycle along the leather value chain, that benefits the totality of the actors involved and, nonetheless, our planet.

4.2 An International Network

It is important to highlight that CFI is not a single case; in fact, there are other POs active globally. An international network was created in 2018, named Carbon Footprint international (https://carbonfootprintinternational.com). The goal is to create a structure, allowing for closer connections between these initiatives, that can support each other facilitating and promoting the exchange of knowledge, lessons learnt, achievements and ideas. On top of this, as mentioned, it is crucial to ensure recognition between initiatives. In fact, many multinational companies work in several markets: there is the necessity to ensure that the achievements communicated in one country are recognized all over the world. The advantage is that CFI, as other POs, are based on the ISO standards, that have a global recognition. In case the rules at the base of the POs are compatible, it is possible to develop some pathways in direction of mutual recognitions. CFI is engaged in promoting MoU (Memorandum of Understanding) agreements between different initiatives. Recently, a MoU was signed between CFI and *The Climate Registry*, and other agreements are in the pipeline. The goal is to move towards a consistent structure, where a mark will be fully recognized in the associated markets. Therefore, for example, the dedicated page of a product can be assessed and is fully recognized in the other markets.

This is also encouraged by the presence of multilateral agreements (MLA) between different accreditation bodies (International Accreditation forum), that ensure an international recognition of the verification activities realized under a certain accreditation scheme in a specific country.

5. Concluding remarks

The paper highlights the quick evolution that the carbon management sector is undergoing and the need for reliable and recognized tools for communicating voluntary efforts embarked by organizations. The ability to support climate claims in a credible way – by using internationally recognized ISO standards – is, and will be even more in the future, a key factor for a successful transition towards decarbonization.

The Italian experience Carbon Footprint Italy, namely the national Programme Operator on carbon management, is drawing an important interest thanks to the innovative application of new technologies to the specific sector, providing a transparent and credible communication Programme. Indeed, it only accepts for registration accredited verified claims. It is an easy-to-use tool, providing unambiguous and unmodifiable information to the public.

It is also important to ensure collaboration and trust along the whole value chain to achieve a real sector transformation, involving and aligning all the interested parties. Transparent communication fosters cooperation along the value chain and boosts market mechanisms based on virtuous actions. It can also provide guidance to organizations undertaking the challenge of decarbonization. Within the leather sector, there is a growing demand for transparency coming from customers, and using carbon footprint labels is one of the most effective ways to address this need.

In parallel, it appears crucial to move towards an international recognition of the existing initiatives. Carbon Footprint International represents the first network at the global level that gathers carbon footprint Programme Operators, fostering the exchange of good practices and experiences. The review shows how this experience is effectively moving the context towards this direction, improving the coordination and the recognition of initiatives.

References

Carbon Footprint International, *The International Network for Carbon Footprint Communication Programmes.* https://carbonfootprintinternational.com

EP 2003, DIRECTIVE 2003/87/EC OF THE EUROPEAN PARLIAMENT AND OF THE COUNCIL, European Parliament, Bruxelles

Evercity 2020, Carbon Footprint Italy and Evercity launched the first blockchain-based solution to store carbon footprint data, available at: https://evercity.medium.com/carbon-footprint-italy-and-evercity-launched-the-first-blockchain-based-solution-to-store-carbon-bd8d8fe7b205

ISO 2000, ISO 14020:2000 Environmental labels and declarations — General principles, International Organization for Standardisation (ISO), Ginevra

ISO 2006, ISO 14025:2006 Environmental labels and declarations — Type III environmental declarations — Principles and procedures, International Organization for Standardisation (ISO), Ginevra

ISO 2017, ISO 14026:2018 Environmental labels and declarations — Principles, requirements and guidelines for communication of footprint information, International Organization for Standardisation (ISO), Ginevra

ISO 2017, ISO/TS 14027:2017 Environmental labels and declarations — Development of product category rules, International Organization for Standardisation (ISO), Ginevra

ISO 2018, ISO 14064-1 Greenhouse gases — Part 1: Specification with guidance at the organization level for quantification and reporting of greenhouse gas emissions and removals, International Organization for Standardisation (ISO), Ginevra

ISO 2019, ISO 14064-3:2019, ISO Greenhouse gases — Part 3: Specification with guidance for the verification and validation of greenhouse gas statements, International Organization for Standardisation (ISO), Ginevra

ISO 2018, ISO 14067:2018, Greenhouse gases — Carbon footprint of products — Requirements and guidelines for quantification, International Organization for Standardisation (ISO), Ginevra

ISO 2020, ISO 14065:2020 General principles and requirements for bodies validating and verifying environmental information, International Organization for Standardisation (ISO), Ginevra

BSI 2010, PAS 2060 Specification for the demonstration of carbon neutrality, British Standards Institution (BSI), London

UNI 2018, ISO 14044:2018 Environmental management – Life cycle assessment – Requirements and guidelines, International Organization for Standardisation (ISO), Ginevra.

UNI 2012, UNI EN 15804:2012 Sostenibilità delle costruzioni - Dichiarazioni ambientali di prodotto - Regole quadro di sviluppo per categoria di prodotto, Ente Nazionale Italiano di Unificazione (UNI), Milan

Vizzoto et al, 2021, *Are Consumers Ready for a LCA-Based Sustainability Facts Panel? Insights From an Online Experiment*, Academy of Management, available at: https://journals.aom.org/doi/10.5465/AMBPP.2021.10555abstract

Whelan et al, 2019, Research: *Actually, Consumers Do Buy Sustainable Products*, Harvard Business Review, available at: https://hbr.org/2019/06/research-actually-consumers-do-buy-sustainable-products

Application of Kinetics Models in Leather Processing

V. Queiroz[1,2], V. Kopp[1], C. Agustini[1], M. Schwaab[2], M. Gutterres[1]

Federal University of Rio Grande do Sul (UFRGS), Chemical Engineering Post-Graduated Program (PPGEQ), Laboratory for Leather and Environmental Studies (LACOURO)[1], Theoretical and Applied Catalytic and Adsorption Process Engineering Laboratory [2], Av. Luiz Englert s/n°, Porto Alegre-RS, Brazil, 90040-040, (51)3308-3954, vania.queiroz@ufrgs.br, mariliz@enq.ufrgs.br

Abstract

Kinetics are present in almost all industrial processes. Its study is the key to clarify the behavior of many chemical systems. Understanding how a reaction occurs, its elementary steps, reaction pathways and kinetic mechanisms, enables the processes improvement. In leather processing, the study of kinetics provides important information about the necessary operating time in beamhouse, tanning, dyeing and fatliquoring stages, so that energy and resources can be saved. In experimental kinetic study, a model can be fitted to the data so that it numerically reproduces the experimental measurements of the process in a satisfactory way. The model fit is important because it allows the optimization of the process without the need for more experiments to be done. In literature, (i) phenomenological kinetic models are found, obtained through the laws of physics and chemistry, as well as the laws of conservation of mass, energy and movement and (ii) empirical models, based on experimental data obtained in the process, which are generally correlated by statistical and mathematical methods, such as regressions. For leather processing, few works use kinetic models, even though a wide variety of studies performing kinetic experiments are found. The aim of this work was to evaluate which kinetic models are used in the leather process, as well as to identify whether the hypotheses made in these models correspond to what occurred experimentally. The most used models are: pseudo-first order, pseudo-second order, Weber-Morris, Cegarra-Puente and modified Cegarra-Puente. These models were applied in the stages of unhairing, tanning and dyeing.

Key words: kinetics, kinetics modeling, leather processing

1 Introduction

The effective utilization of chemical products is an important factor for a sustainable industry as it decreases the effluent volume and the environmental impact (Kanagaraj, 2016). The optimization of the process of leather manufacturing allows the use of the properly amount of chemical needed, reducing the amount of hazardous chemicals and the pollution load of wastewater (Spivak, 2019). This can be achieved using kinetic models that permit simulations and prediction of the evolution of the system under the studied working conditions. With mathematical models, it is possible to scale-up without repeating experiments (Kanagaraj, 2011).

In leather processing, kinetic studies provide important information about the necessary operating time to complete the reaction in beamhouse, tanning and wet-finishing operations so that energy and resources are saved. The nature of the reactants can influence the speed of certain chemical reactions, due to the different process inputs in leather processing, such as raw material (skin/hide - species, race), chemicals (supplier, choice, environmental problems), fashion/trends.

Over the years, a wide variety of kinetics models (Langmuir, Pseudo-first order, Pseudo-second order, Elovich, Crank, Boyd, Bangham, Weber and Morris) have been formulated (Largitte, 2016). In applying such models, it is extremely important to follow the assumptions of the models to obtain a consistent result. In literature, kinetic models are often used in the data processing of wastewater of leather process (Piccin el al., 2012), however few papers use kinetic models for leather processing. The aim of this work was to evaluate which kinetic models have been used to study the steps of unhairing, tanning, and dyeing, as well as to identify whether the hypotheses made in these models correspond to what occurred experimentally. This paper identified the main kinetic models used in leather processing, their hypotheses, conditions in which they are valid and their type (phenomenological or empirical) and verified the utilization of the kinetic models in the literature according to their hypotheses.

2 Methodology

The Scopus and ScienceDirect databases were used to survey the most used kinetic models in leather processing. For the search, the following terms were considered in the title, abstract and keywords: “kinetics” and “leather’’, “kinetics” and “tanning’’, “kinetics” and “dyeing’’, “kinetics” and “unhairing”.

Searches were made in August 2021, and only articles that made the kinetic modeling of leather processing were considered in this work, those that addressed the topic in the treatment of effluents from the tanning process were discarded. The results of the databases research were analyzed to eliminate duplicate articles and exclude inappropriate articles. After this evaluation, the articles remained categorized as to the stage of leather processing in which the kinetics was evaluated (unhairing, tanning, dyeing).

3 Kinetics of leather processing

Leather manufacturing requires several steps, with sequential additions of chemicals, intercalated by washes and mechanical processes (Aquim, 2009). The steps that involve adding chemicals and washing are usually done in wet medium performed in drums. These steps require mass transfer from the bath to the hide. For the unhairing diffusion of chemical products is involved. Tanning and dyeing steps involve the diffusion and fixation of tanning agents and dyes, respectively. It is very important to know the kinetics of the process (Gutterres and Mella, 2015).

3.1 Unhairing

In the unhairing step the hair can be destroyed (hair-burning) or saved (hair saving). In order to preserve the hair, enzymes are used and attack the hair roots and the epidermis (Dettmer et al., 2013). Firstly, to get into the hair roots, the enzymes must diffuse into the hide. The diffusion through the hide matrix takes 70% of the total enzymatic time (Yates, 1969).

3.2 Tanning

The tanning process takes place in two sequential steps: first, the diffusion of the tanning molecules into the hide and the subsequent fixation, in which the tanning agent reacts with the collagen (Fuck, 2008). The reaction mechanism of collagen with tanning agents varies with their chemical characteristics. Carboxylic groups, for example, normally react with mineral tanning agents (chromium). On the other hand, the peptide groups include the carbonyl in which there is interaction by strong hydrogen bonding with vegetable tannins (Gutterres and Silva, 2010). Factors influencing diffusion are the size of the molecule responsible for tanning, solution viscosity, temperature, pH, tanning concentration, mechanical actions, etc. (Faber, 1990). The pH, for example, causes changes in the collagen stereostructure that influence the binding of certain tanning agents (Schroepfer and Meyer, 2015).

The hide is pickled with acidic pH so that the carboxylic groups that make up the collagen structure are protonated taking low affinity with chromium. This low affinity allows a better diffusion of chromium throughout the hide preventing superficial fixation. For diffusion to occur, the hide remains in the drums with chromium salt for about 5 hours, during which time the chromium molecules penetrate the hide matrix due to the difference in chromium concentration (concentration gradient) between the hide's internal matrix and the external matrix of the hide affected in turn by the movement of the drum (Gutterres and Mella, 2014; Sathish, 2016). After the chromium has penetrated the hide, the fixation step can be started. To fix the chromium, it is necessary the carboxylic groups of the collagen structure to be ionized and the chromium salt changes the charge of its complex, going from anionic to cationic. Basifying agents are added to the bath to gradually raise the for pH 2,5 – 3,0 to pH 3.8 - 4.2 with increase of temperature to about 50°C, during the movement of the drum for 1-2 h when the chromium fixes (Gutterres and Mella, 2014).

Tanning with vegetable tannins also takes place in two stages, diffusion and fixation. The conditions for the process to take place are, however, different from those for chrome tanning. Vegetable tannin tanning requires a pre-tanning to open the collagen fibers. Bonding with collagen occurs through the hydrogen bonds of phenolic groups in tanning substances with certain basic groups attached to the polypeptide chain. Fixation occurs at a temperature ranging from 30 to 40°C after the tannin has passed through and at a pH in the range of 3 to 6, to avoid tearing and oxidation of the leather (Auad, 2018, 2020).

3.3 Dyeing

The dyeing process occurs with the distribution of a dye in the solution and its adsorption on the collagen matrix (Püntener, 2000; Kanagaraj et al., 2016). The kinetic behavior of a dye during its transfer from the bath to the leather occurs in at least five steps. They are: 1. Diffusion of dye from the bath to the leather surface; 2. Diffusion through the boundary layer between water / leather; 3. Surface adsorption; 4. Diffusion in the leather matrix (Püntener, 2000) and 5. Fixation (Fuck, 2008).

Diffusion is one of the most important phenomena (Püntener, 2000). Different types of dyes have different rates of adsorption, due the difference in their chemical structure, and this can lead to problems in achieving levelness and shade consistency (Kanth et al., 2009; Kanagaraj et al., 2016). Dye fixation happens due to the chemical reactions between the amine molecule of collagen and the dye molecule (Fuck, 2008). It is not always necessary for the dye to be crossed in all leather cross-section/thickness. Depending on the purpose of the leather article, only superficial dyeing can be done.

Thus, the two steps may or may not occur in sequence. Nonetheless, the high-grade leathers are throughout dyed so that they have the same shade on the cut edge and on the back (flesh side) as on the more compact grain side (Püntener, 2000). Efforts have been made to understand the phenomena involved in the dyeing process, however the physico-chemical mechanisms in leather dyeing are not yet clearly understood. Even if the same procedure and the same chemicals are used, differences in depth and shade level can occur from batch to batch (Muralidharan and Rao 1990; Püntener, 2000).

4 Kinetics models applied to leather processing

The kinetic models applied to leather processing are models usually considered in the kinetic modelling of solid-liquid adsorption. This may happen because the steps in which they are used in leather processing resemble diffusion/adsorption process inside porous adsorbent solids. In unhairing, the chemicals need to diffuse until they reach the hair. In tanning, in a bath containing chromium or vegetable tannin, the tanning agent diffuses into the hide fiber due to concentration gradient, and fixation occurs very similarly to a chemisorption. Similarly, in cross-dyeing, dye molecules in a bath diffuses to the interior of the leather fibers, and the driving force is again a concentration gradient. In literature about leather, most used kinetic models are empirical, based on experimental data obtained in the process, which are generally correlated by statistical and mathematical methods, such as regressions. On the other hand, phenomenological kinetic models are obtained through the laws of physics and chemistry, and the laws of conservation of mass, energy and movement, are rarely used in this application.

The adsorption kinetic models that have been also used in leather processing, are commonly classified as based on reaction control, as the pseudo-first order and pseudo-second order models (Qiu, 2009; Souza, 2017) and based on diffusional control, as the Weber-Morris model (Qiu, 2009). In the reaction control models the adsorption is considered to be the most time-consuming process, so the intraparticle diffusion and external mass transfer can be neglected. Otherwise in models with diffusion control, the diffusion is the rate limiting step. There are still other models used in the leather processing, such as the Cegarra-Puente, modified Cegarra-Puente and Vickerstaff. Some of the most commonly used models in the lather process are presented in the sequence.

4.1 Weber and Morris model

Weber and Morris (1963) have studied the batch adsorption of alkyl benzene sulfonates by activated carbon. The mass transfer into the interior of the particle, characterized by an intraparticle diffusion coefficient, can be the slowest step (Largitte, 2016), as seen in the Equation 1:

$$q_t = k_i\sqrt{t} + C \tag{1}$$

Where q_t is the amount of adsorbent in the adsorbent (mg/g) at time t, that varies linearly as a function of the square root of time t; k_i is constant of the intraparticle diffusion rate and C is a constant that normally is associated to the degree of the boundary effect.

In the work of Schwaab (2017), it was observed that this relationship between the adsorbed concentration versus the square root of time is obtained when a semi-infinite solid is considered, a linear adsorption isotherm and with constant liquid phase concentration. With similar assumptions, it also can be obtained for finite solid, but in this case this relationship in only valid at the beginning of the process, that is, very low time values. For this reason, Weber-Moris model must be used carefully, particularly if the validity of the assumptions cannot be assured.

4.2 Cegarra-Puente

Cegarra and Puente (1967) did the dyeing tests in a bath with a constant capacity. In order to keep the bath concentration constant in the dye bath, an amount greater than that needed to saturate the polyester fiber was used. The process of dyeing polyester fibers with dispersed dyes has the diffusion as the slowest reaction, consequently, is the one which sets the rate of the process.

The Cegarra-Puente equation is based on heterogeneous reactions delayed by the reaction chemicals where a desorption rate appears. This empirical equation relates the rate of dye absorption to the amount of dye absorbed by the substrate at any time during the dyeing process (Cegarra and Puente, 1967; Kanth et al., 2009). Equation 3 was proposed by Cegarra and Puente.

$$\ln\left[1-\frac{C_t^2}{C_\infty^2}\right]=-Kt \tag{2}$$

Where C_∞ is the concentration in the adsorbate at the equilibrium, C_t is the concentration of adsorbate in solution at time t and K is the constant proposed by Cegarra and Puente.

4.3 Modified Cegarra-Puente / Cegarra-Puente-Valdeperas

Some modifications were proposed to the Cegarra-Puente model, one of them was carried out in the doctoral thesis of J. Valdeperas, in which it was assumed that the direct velocity, in relation to what it would have in a bath of constant concentration, would decrease over time. Valldeperas suggested that the forward speed be proportional to a power of time, which would also affect the speed of the inverse reaction. The equation is adapted to tannin baths with varying exhaustion, introducing the appropriate modification, where 'a' is the coefficient depending on exhaustion (Sánchez, 1981). The modified equation is presented in Equation 3:

$$\ln\left[-\ln\left[1-\frac{C_t^2}{C_\infty^2}\right]\right]=a\ln t+a\ln K \tag{3}$$

Another way to write Equation 3 is as shown below in Equation 4:

$$\frac{C_t}{C_\infty}=\left[1-e^{-(kt)^a}\right]^{1/2} \tag{4}$$

This equation allows the calculation of *Ct* for different times, knowing the velocity constant and the exponent. Where $C\infty$ is the concentration in the adsorbate at the equilibrium, C is the concentration of adsorbate in solution at the moment t, K is the constant proposed by Cegarra and Puente, t is the time, and a is the coefficient depending on exhaustion.

4.4 Vickerstaff

Using a diffusion model, Vickerstaff developed an approximation to Crank's equation to describe dyeing (Jian et al., 2011), as shown in Equation (5).

$$C_t = 2C_\infty \sqrt{\frac{Dt}{\pi}} \tag{5}$$

Where Ct was the concentration of released total protein, $C\infty$ is the concentration at the equilibrium, D was the diffusion coefficient and t was the time.

4.5 Pseudo-first-order

The pseudo-first-order is an empirical model proposed by Lagergren (1898) to describe the kinetic process of adsorption in the liquid-solid phase of oxalic acid and malonic acid on charcoal. It is commonly used for liquid-solid adsorption systems. The pseudo-first-order model describes the adsorption kinetics of a species on an adsorbent particle by the first order differential equation (6).

$$\frac{dq_t}{dt} = k_1 (q_e - q_t) \tag{6}$$

According to this model, the adsorption capacity is proportional to the "distance to equilibrium" which is expressed by the difference between the equilibrium concentration and the amount adsorbed at time t (Rodrigues and Silva, 2016). Its integrated form is commonly used, by integrating this differential equation Eq. (7) for the boundary conditions: qt = 0 at = 0 and qt = qt at = t, the following equation can be obtained (Benjelloun, et al. 2021):

$$q_t = q_e \left(1 - k_1 e^{-k_1 t}\right) \tag{7}$$

Where qe is the amount of adsorbate in the equilibrium adsorbent (mg/g); qt is the amount of adsorbent in the adsorbent at time t (mg/g); $k1$ is Lagergren's first order constant rate; and t is the contact time (min).

4.6 Pseudo-second-order

The pseudo-second-order model was proposed by Ho (1998) to describe a kinetic process of adsorption of divalent metal ions onto peat. The model's assumptions were that the adsorption can be of second order and the rate limiting step can be chemical adsorption involving valence forces through the sharing or exchange of electrons between the peat and the divalent metal ions. Furthermore, the adsorption follows the Langmuir equation. Kinetics can be expressed by the following second order differential equation Eq. (8)

$$\frac{dq_t}{dt} = k_2 \left(q_e - q_t\right)^2 \tag{8}$$

Its integrated form is commonly used, by integrating this differential equation Eq. (8) for the boundary conditions: qt = 0 at = 0 and qt = qt at = t, the mathematical equation of the following pseudo-second order kinetic model Eq. (9) was obtained:

$$q_t = \frac{q_e^2 k_2 t}{1 + q_e k_2 t} \tag{9}$$

Where *qe* is the amount of adsorbate in the equilibrium adsorbent (mg/g); *qt* is the amount of adsorbent in the adsorbent at time *t* (mg/g); K_2 is the pseudo-second-order constant rate; and *t* is the contact time (min).

Table 1 presents the main hypotheses made in the deduction of the models and which step is the limiting of the considered model are presented. In this compilation of the kinetic models is indicated its use in different steps of leather processing as found in the literature.

Table 1: Kinetic models applied in leather processing referenced in the literature.

Kinetic Model	Equation	Hypothesis	Reference
Cegarra and Puente	$\ln\left[1-\frac{C_t^2}{C_\infty^2}\right]=-Kt$ *	Bulk phase concentration remains constant Process reversibility	Kanth 2009 Kanth 2010
Modified Cegarra and Puente	$\ln\left[-\ln\left[1-\frac{C_t^2}{C_\infty^2}\right]\right]=a\ln t+a\ln K$ *	Varying exhaustion	Kanth 2009 Kanth 2010
Weber-Morris	$q_t=k_i\sqrt{t}+C$	Bath concentration is constant The adsorption kinetics is instantaneous The equilibrium isotherm is linear	Kanth 2009 Sathish 2019
Pseudo-first order	$\frac{dq_t}{dt}=k_1\left(q_e-q_t\right)$ $\ln\left(q_e-q_t\right)=\ln q_e-k_1t$ *	The adsorption driving force is the difference between the solid phase concentration at any time and at equilibrium	Sathish 2019
Pseudo-second order	$\frac{dq_t}{dt}=k_2\left(q_e-q_t\right)^2$ $q_t=\frac{q_e^2k_2t}{1+q_ek_2t}$	The adsorption driving force is the square of the difference between the solid phase concentration at any time and at equilibrium	Sathish 2019 Kanagaraj 2011 Kanagaraj 2016
Vickerstaff	$Ct = 2C\sqrt{\frac{Dt}{\pi}}$	Infinite bath	Jian 2011

*linearized equation

5 Results and discussions

Table 2 presents the application of the kinetic models to each leather processing step. In the literature, the models used are mostly empirical (Cegarra-Puente, modified Cegarra-Puente, Pseudo-first order and Pseudo-second order). They are easy to use, but are subjected to operational and process conditions in which they were deduced and, therefore, cannot be used to describe the system with conditions different from those determined for its adjustment, which limits its application for prevision.

It can be seen that kinetics is studied mainly in the tanning and dyeing steps, steps that involve the incorporation of materials into the hide. These two steps of processing lead to pollution of effluent. In the tanning step, not all chromium salt is fixed in the leather, and the remaining chromium go to the effluent (Kopp 2021). At dyeing, 1 to 5% of the dye applied in the leather dyeing remains in the effluent (Fuck et al. 2018) and the effluents from dyeing are difficult to treat by conventional methods (Ortiz-Monsalve et al. 2019).

The influence of pseudo-first and pseudo-second order and Weber-Morris models when it comes to kinetics is notorious, being present in most works, even though these models are very restricted and depend on specific conditions for their application.

The pseudo-first order, pseudo-second order, Cegarra-Puente and Cegarra-Puente modified models used in the analyzed articles were used in their linearized form. The correlation coefficient or sum of squares was used as an index of quality in fitting the data to the model. Transforming the nonlinear equation into a linear form involves assumptions, and the obtained results can be very undermined. This leads to data distortion and, consequently, results that violate the standard least squares assumption of variance and normality. The linearization of the pseudo-second model results in more than five different linear equations as a result one can get different parameters when using different forms of the same model for a given process. Using the non-linear form of the function results in the error being distributed along the curve (Chowdhury and Saha, 2011; Kajjumba, et al 2018).

Unhairing

Jian (2011) used enzymes in the unhairing step. The enzyme protease did little harm on skin at the first 3 h, and after this time it degraded collagen and resulted in serious injury to skin. Using a different enzyme extract of *Bacillus subtilis* BLBc, samples started to show some damages to the hide grains above 6 h of reaction (Dettmer et al., 2013). The kinetics of released collagen protein indicates that the injury to skin took place in 3-6 h. Therefore, it's necessary to control the time of protease unhairing within an appropriate limit. The good fit of Vickerstaff model suggests that enzymatic unhairing by protease is a diffusion-controlled process at the initial stage. Where Vickerstaff can be used only in a short time. An alternative was studied by Andreoli et. al. (2015) in which enzymatic-oxidative hair removal with enzyme extracted from two strains of Bacillus subtilis (BLBc 11 and BLBc 17) and hydrogen peroxide was used, obtained as a result a hair removal that does not promote degradation for hair removal.

Tanning

As seen before, the tanning step, with vegetable tannins or chromium, takes place in two well-defined steps of diffusion and fixation. So, the kinetic study must be carried out taking into account these separate steps that do not occur simultaneously. The papers studied the diffusion stage, which is very important due to the crossing of the chemical to guarantee the stabilization of the hide, in the case of tanning. So, only kinetic models that are based on diffusion should be used. However, pseudo-first-order and pseudo-second-order are based on reaction (Wang 2020) and were used.

Sathish (2019) studied chrome tanning using Weber-Morris, pseudo-first-order and pseudo-second-order. The best fit model has been chosen based on a linear regression coefficient (R^2) and squared sum of errors (SSE) values and also by comparing the experimentally measured *qe* with best fitted theoretically calculated *qe*. The results indicate that the Weber-Morris was the best kinetic model. Thus, the intraparticle diffusion is controlling the sorption of Cr (III) from bulk solution. However, the process is "complex" with more than one mechanism limiting the rate of sorption. Kanth (2010) based the result on the shape of the isotherm and the R^2 values to support the modified Cegarra-Puente model for absorption of tannins.

Dyeing

Kanth (2009) reported the application of enzyme acid protease for the improvement of the exhaustion of dyes. The shape of the isotherm and the R values support the modified Cegarra-Puente model for adsorption of both the dyes. Kanagaraj (2016) used a copolymer from trimming waste hydrolysate that has been developed using starch and poly vinyl alcohol as high performance auxiliary in dyeing. The carboxylic acid groups present in the copolymer are responsible for enhancing the adsorption of dye. Studies on kinetics of dye-uptake suggest that the dye molecules approach the collagen fiber, get adsorbed, slowly diffuse through pores and get fixed with the functional groups of collagen by electrostatic forces. The rate of absorption of dye inside the porous leather was represented by pseudo-second-order kinetics. Kanagaraj (2011) used a nanoparticle polymer to enhance the dye uptake. The reason for the improved uptake of dye may be due to the presence of carboxyl groups that play as the main functional group of interaction with dye compounds. The rate of mass absorption inside the porous leather was represented by pseudo-second-order kinetics.

Table 2: Application of kinetic models in leather processing report in the literature.

Chemical product	Process conditions	Kinetic model	According to the kinetic model?*	References
		Unhairing		
Enzymes	Enzyme: 150% (w/w) Temperature: 35°C Rotation: 8 rpm	Vickerstaff	Yes	Jian 2011
		Tanning		
Chrome	Basic chromium sulfate: 5% w/w Time: 15 to 400 min Temperature: 37°C	Weber Morris* Pseudo-first-order Pseudo-second-order	Yes	Sathish 2019
Vegetable tannin	Wattle (tannin): 20% Time: 5 to 120 min Temperature: 25, 30, 35 and 40°C	Weber Morris (Parabolic) Cegarra- Puente Modified Cegarra-Puente*	Yes	Kanth 2010
		Dyeing		
Acid Black 194 and Acid Black 210	pH: 4.5 50°C	Weber-Morris Cegarra- Puente Modified Cegarra-Puente*	Yes	Kanth 2009
Acid brown MFR	different temperature, pH, duration and concentration of dye	Pseudo-second-order	No	Kanagaraj 2016
Acid Black 20470	Dye: 1% or 2% (based on leather weight) Time: 1 h, 50, 60, 70, and 80h pH: 5.6	Pseudo-second-order	No	Kanagaraj 2011

6 Conclusions

This study identified few works focused on the study of leather processing kinetics, in contrast to the large amount that is found for the kinetics of treatment of wastewater from the leather industry. The few works found in the literature used empirical models or simplification of phenomenological models. These models have as main limitation the fact that they are only valid under specific conditions in which they were deduced, or in the case of simplified models under the conditions considered for simplification. The most used models (pseudo-first order, pseudo-second order, Cegarra-Puente and modified Cegarra-Puente) are usually used in their linearized forms. However, it is not very appropriate to use the linear method to determine kinetic parameters, because transforming a non-linear kinetic model into a linearized form tends to change the error distribution and therefore distort the parameters. Thus, it is advised that more research in this regard should be carried out, in order to better understand the kinetics of the unhairing, tanning and dyeing steps. Furthermore, the use of phenomenological models should be encouraged.

7 Acknowledgements

The authors would like to thank the CAPES for the financial aid received.

References

Andrioli, E., Petry, L., Gutterres, M. (2015): *Environmentally friendly hide unhairing: Enzymatic-oxidative unhairing as an alternative to use of lime and sodium sulfide.* Process Safety and Environmental Protection, 93, 9-17.

Aquim, P. M. D., Gutterres, M., Trierweiler, J. O., (2009): *Assessment of water management in tanneries: State of Rio Grande do Sul case study.* Proceedings of the XXX Congress of the IULTCS, China, 2009. (30.: 2009 out. 11-14: Pequim, China)

Auad, P., (2018): *Composição química de taninos vegetais, curtimento e propriedades nos couros. Master thesis.* University Federal of Rio Grande do Sul.

Auad, P., Spier, F., and Gutterres, M. (2020): *Vegetable tannin composition and its association with the leather tanning effect.* Chemical Engineering Communications, 207, 722-732.

Benjelloun, M., Miyah, Y., Evrendilek, G. A., Zerrouq, F., and Lairini, S. (2021): *Recent Advances in Adsorption Kinetic Models: Their Application to Dye Types*. Arabian Journal of Chemistry, 103031.

Chowdhury, S., and Saha, P., (2011): *Adsorption kinetic modeling of safranin onto rice husk biomatrix using pseudo-first-and pseudo-second-order kinetic models: Comparison of linear and non-linear methods.* CLEAN–Soil, Air, Water., 39, 274-282.

Dettmer, A., Cavalli, E., Ayub, M., Gutterres, M (2013). *Environmentally friendly hide unhairing: enzymatic hide processing for the replacement of sodium sulfide and delimig*. Journal of Cleaner Production, 47, 11-18.

Faber, K., (1990): *Bibliotek des Leders Band 3: Gerbmittel.* Frankfurt: Gerbund und Nachgerbund., *30, 95.*

Fuck, W. F., (2008): *Influência do teor de cromo e do tipo de óleo na formação de cromo hexavalente no couro. Master thesis.* University Federal of Rio Grande do Sul.

Fuck, W. F., Lopes, F. C., Brandelli, A., Gutterres, M., (2018): *Screening of Natural Dyes from Filamentous Fungi and Leather Dyeing with Monascus purpureus Extract.* Journal of the Society of Leather Technologists and Chemists., 102, 69-74.

Gutterres, M., and da Silva, I., (2010): *Leather retanning with hydrolyzed protein.* Journal of the American Leather Chemists Association., 105, 06.

Gutterres, M., and Mella, B., (2014): *Chromium in tannery wastewater.* Heavy Metals In Water: Presence, Removal and Safety., 315, 315-344.

Ho, Y. S., and McKay, G., (1998): *A comparison of chemisorption kinetic models applied to pollutant removal on various sorbents.* Process safety and environmental protection., *76*, 332-340.

Jian, S., Wenyi, T., Wuyong, C., (2011): *Kinetics of enzymatic unhairing by protease in leather industry.* Journal of Cleaner Production., 19, 325-331.

Kajjumba, G. W., Emik, S., Öngen, A., Özcan, H. K., Aydın, S., (2018): *Modelling of adsorption kinetic processes—errors, theory and application.* Advanced sorption process applications., 187-206.

Kanagaraj, J., and Panda, R. C., (2011): *Modeling of dye uptake rate, related interactions, and binding energy estimation in leather matrix using protein based nanoparticle polymer.* Industrial & engineering chemistry research., 50, 12400-12408.

Kanagaraj, J., Panda, R. C., Senthilvelan, T., Gupta, S., (2016): *Cleaner approach in leather dyeing using graft copolymer as high performance auxiliary: related kinetics and mechanism.* Journal of Cleaner Production., 112, 4863-4878.

Kanth, S. V., Venba, R., Jayakumar, G. C., Chandrababu, N. K., (2009): *Kinetics of leather dyeing pretreated with enzymes: Role of acid protease.* Bioresource technology, 100, 2430-2435.

Kanth, S., Kannan, P., Usharani, N., Jayakumar, G., Deepa, S., Chandrasekaran, B., (2010). *Studies on the use of enzymes in tanning process: part II. kinetics of vegetable tanning process.* Journal of the American Leather Chemists Association., 105, 16-24.

Kopp, V., Agustini, C., Gutterres. M., Dos Santos, J.H.Z., (2021). *Nanomaterials to help eco-friendly leather processing*. Environ Sci Pollut Res Int., 40, 55905-55914.

Largitte, L., and Pasquier, R., (2016): *A review of the kinetics adsorption models and their application to the adsorption of lead by an activated carbon.* Chemical Engineering Research and Design, 109, 495-504.

Muralidharan, D., and Rao, V. S., (1990): *Studies on the chemical and physical interactions between collagen and leather dyes.* Journal of the Society of Dyers and Colourists., 106, 139-142.

Ortiz-Monsalve, S., Valente, P., Poll, E., Jaramillo-García, V., Pegas Henriques, J.A., Gutterres, M., (2019): *Biodecolourization and biodetoxification of dye-containing wastewaters from leather dyeing by the native fungal strain Trametes villosa SCS-10*. BiochemEng J., 141, 19–28.

Piccin, J. S., Gomes, C. S., Feris, L. A., Gutterres, M., (2012): *Kinetics and isotherms of leather dye adsorption by tannery solid waste.* Chemical Engineering Journal., 183, 30-38.

Püntener, A. G., (2000): *Leather dyes.* In: H.S. Freeman and A.T. Peters (eds). In Colorants for Non-Textile Applications, 478-557, (Elsevier Science)

Qiu, H., Lv, L., Pan, B. C., Zhang, Q. J., Zhang, W. M., Zhang, Q. X., (2009): *Critical review in adsorption kinetic models.* Journal of Zhejiang University-Science A., 10, 716-724.

Rodrigues, A. E., and Silva, C. M., (2016): *What's wrong with Lagergreen pseudo first order model for adsorption kinetics?*. Chemical Engineering Journal., 306, 1138-1142.

Sánchez, J. C., Puente, P., Valldeperas, J., (1981): *Fundamentos científicos y aplicados de la tintura de materias textiles*. Universidad Politécnica de Barcelona.

Sathish, M., Dhathathreyan, A., Rao, J. R., (2019): *Ultraefficient tanning process: Role of mass transfer efficiency and sorption kinetics of Cr (III) in leather processing. ACS Sustainable Chemistry & Engineering.*, 7, 3875-3882.

Schroepfer, M., and Meyer, M., (2015): *Investigations towards the binding mechanisms of vegetable tanning agents to collagen.* Proceedings of the XXXIII Congress of the IULTCS, Brazil, 2015.

Souza, P. R., Dotto, G. L., Salau, N. P. G., (2017): *Detailed numerical solution of pore volume and surface diffusion model in adsorption systems.* Chemical Engineering Research and Design., 122, 298-307.

Souza, F. D. R., Benvenuti, J., Meyer, M., Wulf, H., Klüver, E., and Gutterres, M. (2020): *Extraction of keratin from unhairing of bovine hide.* Chemical Engineering Communications, 1-9.

Spivak, S., and Kantor, O., (2019). *Mathematical Modeling and Optimization of Chemical and Technological Processes.* In: Proceedings of the 1st International Conference on Control Systems, Mathematical Modelling, Automation and Energy Efficiency (SUMMA), 35-40, Russia.

Wang, J., and Guo, X., (2020): *Adsorption kinetic models: Physical meanings, applications, and solving methods.* Journal of Hazardous Materials., 390, 122-156.

Xiao, Y., Azaiez, J., Hill, J. M., (2018): *Erroneous application of pseudo-second-order adsorption kinetics model: ignored assumptions and spurious correlations.* Industrial & Engineering Chemistry Research., 57, 2705-2709.

Yates, J.R., (1969). S*tudies in depilation: part IX. Effect of skin thickness and diffusion on the rate of depilation of sheepskins*. J. Am. Leather Chem. Assoc. 64, 71-81.

Melamine Ethoxylates: Novel Formaldehyde free Resin Re-Tanning Agents

Jochen Ammenn

Stahl, Benzstraße 11, 70771 Leinfelden-Echterdingen; Germany; +49 174 1878217

jochen.ammenn@stahl.com

Abstract. Melamine urea formaldehyde condensates or resin re-tanning agents are commonly used in the leather industry. They give rise to homogeneous filling, are easily and widely applicable, but generally contribute to free formaldehyde, measurable in emission or extraction methods from leather. Many projects to resolve this issue have been carried out. The weakness of the resin re-tanning is the reversibility of the bond formation between formaldehyde and two amino groups. The formaldehyde forms a methylene group after condensation with two amino groups and releases water. This methylene group can be hydrolyzed under acidic conditions to release formaldehyde again.

Melamine ethoxylates were found to be a formaldehyde free, stable, and easily applicable alternative. Besides the facts that they are made without formaldehyde, it proved to be possible to incorporate renewable components into the synthesis, giving rise to an even more sustainable product. Another sustainable aspect of this chemistry is the delivery of the potential product as a concentrated liquid versus most resin re-tanning agents being placed into the market as spray dried powders to remove the water in which they are made, consuming significant amounts of energy.

Melamine ethoxylates show advantages in comparison to a classical resin re-tanning agent in application on leather. An insight into the scope of application will be given, showing that melamine ethoxylates can be considered a formaldehyde free replacement for resin re-tanning agents.

1 Introduction

Within the leather making process the re-tanning step has the role to determine main leather article properties like fullness, softness, grain tightness, fastness properties, and tear strength. Major classes of re-tanning agents are phenol formaldehyde condensates and melamine urea formaldehyde condensates. Depending on the chemistry, rest monomeric phenol, formaldehyde or bisphenol S can be a problem with the phenol formaldehyde condensates (1), also known as syntans. Besides melamine urea formaldehyde also urea formaldehyde condensates are applied in re-tanning agents, representing another class, commonly referred to as resins. Both, melamine urea formaldehyde and urea formaldehyde condensates have similar properties on leather and face the same problem: slow release of formaldehyde. This process will be closer described.

Melamine ethoxylates were developed as an alternative to melamine urea formaldehyde condensates, resolving this problem by avoiding the use of formaldehyde. In their formulation other sustainable aspects were incorporated like partially renewable components and moving away from the energy consuming spray drying process. Application results on a widespread of leather articles and details about fastness properties will be given.

2 Materials and Methods

2.1 Melamine urea formaldehyde condensates

Urea formaldehyde condensates were discovered in 1884 (2). Since the 1930'urea formaldehyde condensates were used as impregnating resins in the production of chipboards. This helped to make the wood industry far more sustainable, since the huge amounts of waste in sawing of trees could be converted to chipboards, which were considered products of value made from waste. Later melamine formaldehyde condensates and melamine urea formaldehyde were developed, giving additional characteristics to chipboards like fire resisting properties.

Melamine formaldehyde condensates we first applied on leather in 1942 as tanning agents in form of a formaldehyde releaser (3). For application as re-tanning agents in leather the water solubility had to be improved. This issue was the same for both melamine urea formaldehyde and urea formaldehyde condensates. We will focus on melamine urea formaldehyde condensates, keeping in mind that problems of these two similar chemistries and approaches to resolve these problems are parallel.

Melamine urea formaldehyde condensates are generally made under moderate basic or acidic conditions in water (5), as can be seen in figure 1. The chain length, reflected in the number "n", is depended mainly on the amount of formaldehyde, but also on other reaction conditions as time and temperature. The bonds deriving from formaldehyde are marked in red.

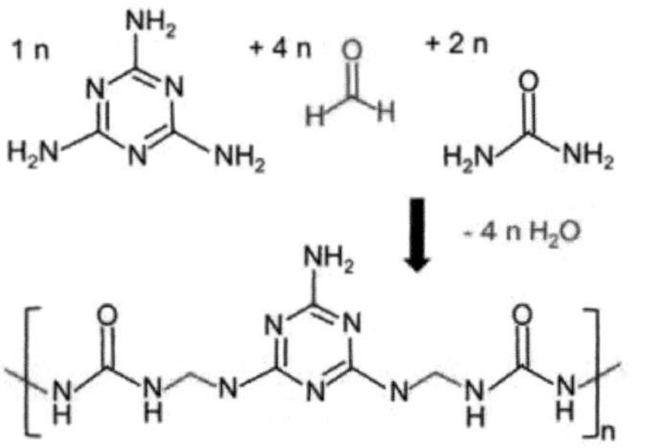

Fig. 1. Synthesis of melamine urea formaldehyde condensates

The issue of improving water solubility could be resolved by adding sulfites to the condensation (4), after having formed a poly condensate with more equivalents of formaldehyde. The sulfite condenses with amino methylol groups deriving from the excess of formaldehyde, releasing water and forming a sulfonic acid group attached to the poly condensate via a methylene group. The resulting structure is depicted in figure 2.

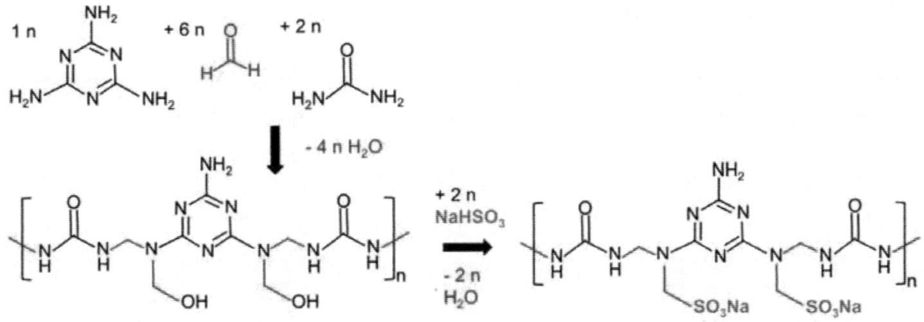

Fig. 2. Water soluble melamine urea formaldehyde condensates

In a basic pH-range the newly formed methylene bridges between two amino group deriving from formaldehyde are quite stable.
The re-tanning process is commonly carried out in floats at acidic pH ranging from pH = 3 to pH = 6. In this pH range the melamine urea formaldehyde condensates are not stable and slowly begin to hydrolyze the methylene, bridging two amino functionalities and releasing formaldehyde. This problem is inherent to this class of chemistry. The speed of hydrolysis is increased by lowering the pH, increasing the amount of acid or increasing the temperature (6). The hydrolysis is depicted in figure 3.

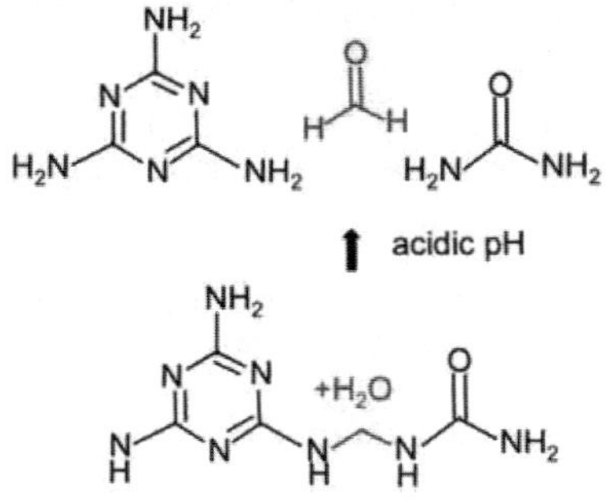

Fig. 3. Hydrolysis of a melamine urea formaldehyde condensate

The release of formaldehyde can be quantified in emission or extraction assays. In the emission test according to UNI EN 27587 a flow nitrogen is blown through a test tube with the pure product as powder product or liquid product. Emitted formaldehyde is carried with the nitrogen flow and directed into a cartridge where the formaldehyde is derivatized. The derivatized formaldehyde can be washed from the cartridge and quantified via High Pressure Liquid Chromatography (HPLC). This method is pH neutral. The pH of the test sample will not change during the test. If the melamine urea formaldehyde condensate is adjusted to a basic pH, low values of formaldehyde will be obtained with this method.
In the extraction assay UNI EN 17226 leathers samples can be analysed for formaldehyde. This is especially interesting for leathers made with formaldehyde condensates. The leathers are cut into pieces, extracted with warm water for a defined time. The aqueous solution is derivatized under acidic condition and the derivatized solution is submitted to High Pressure Liquid Chromatography for quantification. Due to the acidic pH in solution, the melamine urea formaldehyde condensates are prone to partially hydrolize. The acidity originates from strong acids, commonly used to establish the reaction conditions for derivatization with 2,4-di-nitro-phenyl-hydrazine (7).

Consequently, higher values of formaldehyde will be obtained with this method for melamine urea formaldehyde condensates.
From an application view in the leather industry the melamine urea formaldehyde condensates give rise to homogeneous filling, medium softness without compromising on grain tightness, if used in re-tanning.

2.2 Melamine ethoxylates

Various attempts have been made in the past to overcome the formaldehyde release from melamine urea formaldehyde condensates. We developed melamine ethoxylates as a potential alternative to melamine urea formaldehyde condensates and urea formaldehyde condensates. One of the first decisions we had to make was to decide on the chemicals to use. Having had the choice between ethylene oxide and ethylene carbonate, we clearly favored ethylene oxide over ethylene carbonate for sustainability reasons, due to the fact that in a reaction with ethylene carbonate one equivalent of carbon dioxide is released. The first hurdle resulting from that decision was to enable the solid and quite unreactive melamine to react with gaseous ethylene oxide. This issue could be resolved by dissolving the melamine in a solvent of choice. In order to comply with sustainable principles, we decided for a renewable solvent and identified glycerine out of a screening. Glycerine is widely available for instance from the production of bio diesel. We then screened the reaction conditions of the ethoxylation (8). A typical example was carried out like this:

24 g of melamine were dissolved in 24 g glycerine in a 300 mL steel autoclave and warmed to 160°c under stirring. To this solution 192 g ethylene oxide were dosed within 6 hours. The reaction was allowed to proceed at this temperature for 3 hours until the pressure became constant. The pressure was released and a vacuum of 15 mbar was applied to the autoclave to remove volatile components. 235 g of a yellow liquid with low viscosity (935 m Pa s) was obtained.

The minimum requirement for a successful reaction was the complete reaction of the melamine in order to achieve a final product without toxic components, since melamine bears the hazardous label H 361 f, meaning it is suspected to have an impact on fertility. The melamine ethoxylate is a liquid with low viscosity that contains no water. It is readily soluble in water and therefore could be easily applied on leather. The synthesis of melamine ethoxylates is depicted in figure 3.

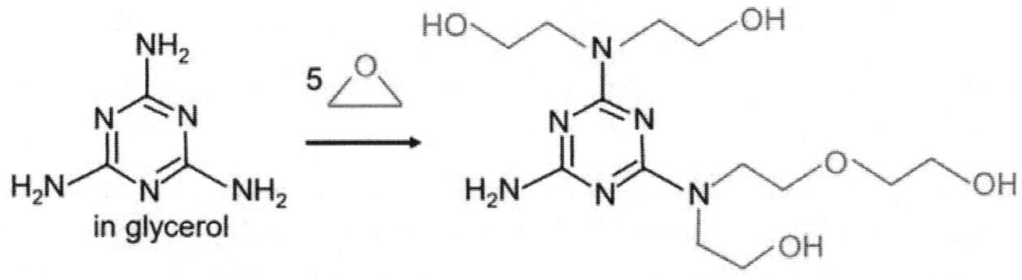

Fig. 3. Synthesis of melamine ethoxylates

2.3 Re-tanning of leather

The re-tanning was carried out on bovine wet blue or wet white with two millimeters strength after neutralization with eight percent active material in float lengths of hundred to two hundred percent, based on wet blue or wet white weight. Subsequently, the leathers were treated with one percent of a standard brown dye and with two percent of a standard fatliquor in a separate float. After re-tanning the leathers were dried and stacked. The comparison was run with a standard melamine urea formaldehyde resin and a poly ethylene glycol, subjected to the same procedure. The lightfastness was determined via Xeno test in a Suntester according to DIN EN ISO 105-B06. The softness can be quantified according to DIN EN ISO 17235, or via haptics. Grain tightness and fullness were evaluated via haptics.

3 Results and Discussion

The screening to identify optimized ethoxylation conditions for melamine is summarized in table 1:

Table 1. Results of re-tanning of bovine wet blue with melamine ethoxylates with varying chain length in comparison to a standard melamine urea formaldehyde condensate and a short chained poly ethylene glycol: + = better; 0 = similar; - = worse

Product	Number of ethylene oxide equivalents / 1 equivalent of melamine	Leather properties compared to standard melamine urea formaldehyde	Leather properties compared to poly ethylene glycol (PEG)
Melamine Ethoxylate 3 EO	3	-	+
Melamine Ethoxylate 5 EO	5	0	+
Melamine Ethoxylate 7 EO	7	-	+

Melamine Ethoxylate 10 EO	10	-	0

As can be seen in table 1, with a minimum of three equivalents of ethylene oxide per one equivalent of melamine, we were able to see a re-tanning effect. With five equivalents ethylene oxide we observed on optimum that did not improve raising the equivalents ethylene oxide to seven or more, if compared to a melamine urea formaldehyde resin. From ten equivalents ethylene oxide and onwards, we still saw a re-tanning effect, but this effect could also be achieved by re-tanning with poly ethylene glycol (PEG). Poly ethylene glycols are made from pure ethylene oxide, are widely available and show only weak re-tanning effects. Goal of our effort was to obtain superior results to re-tanning with poly ethylene glycol and similar effect to the standard melamine urea formaldehyde condensate. This was achieved with the sample deriving from five equivalents ethylene oxide. We scaled this composition up and tested this "pilot batch" in further application tests.

Table 2. Results of re-tanning of bovine wet blue or wet white with melamine ethoxylates in a mixed recipe compared to the same recipe with a standard melamine urea formaldehyde condensate: + = better; 0 = similar; - = worse

	Shoe Upper	**Shoe waterproof**	**Boot Nappa**	**Automotive Wet blue**	**Upholstery**	**Automotive wet white**
Colour strength	+	+	+	+	+	+
Filling	-	-	-	-	-	-
Softness	0	0	+	+	+	0
Grain tightness	0	0	0	0	0	0

As can be seen in table 2 in six different leather articles made with re-tanning recipies typical for these articles, the melamine ethoxylate gave superior results in colour strength compared to a standard melamine urea formaldehyde condensate. Three leather articles resulting from re-tanning with melamine ethoxylate showed superior softness. Grain tightness was similar for both chemistries. Observed fullness was less pronounces for leather articles made with the melamine ethoxylates.

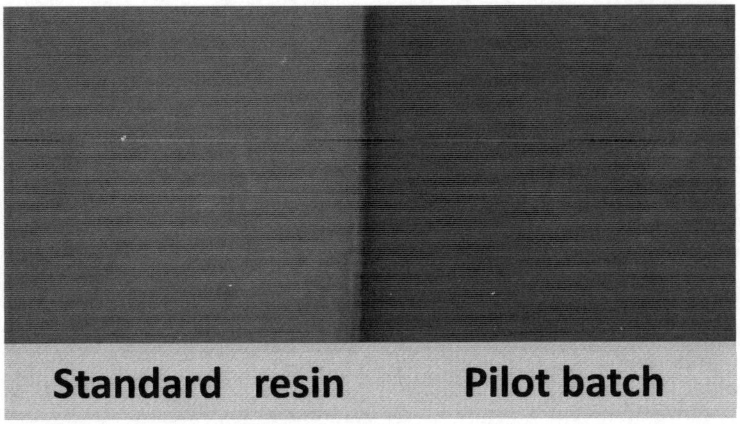

Fig. 4. Re-tanning example of melamine urea formaldehyde resin (left) compared to the pilot batch of melamine ethoxylates

An example of the colour strength advantage of the melamine ethoxylate is shown in figure 4. Results of lightfastness tests, determined via Xeno test in a Suntester according to DIN EN ISO 105-B06, gave clearly better results of the melamine ethoxylate than for the melamine urea formaldehyde condensate.

The melamine ethoxylate is a liquid with low viscosity that contains no water. It is readily mixable with water. This is a sustainable advantage over classical melamine urea formaldehyde condensates, which are made in water and are mostly subjected to spray drying, affording huge amounts of energy to remove the water in order to obtain a concentrated product. Another aspect of improved sustainability is the glycerine inside the product, giving rise to partially renewable starting materials, in contrast to entirely oil-based melamine urea formaldehyde condensates. Overall, we identified melamine ethoxylates as a sustainable alternative to melamine urea formaldehyde condensates.

References

1. *Ammenn, Jochen; Huebsch, Corinna; Schilling, Emil; Dannheim, Bernd;* ***Chemistry of Syntans and their Influence on Leather Quality****, JALCA, 110, 349–354, 2015.*
2. *Tollens, Bernhard;* ***Ueber einige Derivate des Formaldehydes****, Chem. Ber., 17, 653-659, 1884.*
3. *Dawson, William;* ***Tanning Agents****, US 2316741, 1943, 3p.*
4. *Lach, Dietrich; Streicher, Rolf; Bolz, Gerhard;* ***Verfahren zur Herstellung von wasserlöslichen oder selbstdispergierenden Harzgerbstoffen für die Nachgerbung von Leder und ihre Verwendung****, EP 0 063 319, 1982, 10p.*
5. *Shriver, Donald et al.;* ***Production of melamine-urea-formaldehyde concentrates****, US 3.458.464, 1969, 3p.*
6. *Liu, Ming; Wang, Yan; Wu, Yiang; Wan, Hui;* ***Hydrolysis and recycling of urea formaldehyde resin residues****, J. of Harzardous Materials, 355, 96-103, 2018.*
7. *Brady, Ocsar; Elsmie, Gladys;* ***The use of 2:4-dinitrophenylhydrazine as a reagent for aldehydes and ketones****, Analyst., 51(599), 77-78, 1926.*
8. *Ammenn, Jochen; Kunst, Andreas;* ***Process for producing leather****, WO 2011 158168, 13 p.*

Acknowledgements

I wish to express my gratefulness to my collaboration partner Andreas Kunst (BASF, Ludwigshafen, Germany).

Phenolic monomers in aromatic syntans and their influence on leather: (Pending) Restricted compounds

Ramakrishna Prasad Muppa*, Mark Wijland, Petra Berends, and Kenan Özcan

Smit & zoon, Nijverheidslaan 48, 1382LK Weesp, The Netherlands

*Corresponding author, Email: prasad.muppa@smitzoon.com

Aromatic syntans are widely used in the (re)tanning of leathers. Phenolic monomers such as phenol, phenolsulfonic acid, bisphenol S and bisphenol F are common compounds in aromatic syntans. These compounds are unreacted molecules that are either starting materials or intermediate molecules formed during the manufacturing process. Due to the toxic nature of these compounds, both the regulatory authorities and the leather industry are displaying an increasing interest in these residual phenolic monomers. There are various regulatory proposals in discussion on the amount of phenolic monomers (e.g., a restriction proposal on bisphenols is pending at ECHA.) When approved, it will lead to a combined content restriction of bisphenols in leather articles of 200mg/kg (or 0.02% or 200 ppm). Similarly, bisphenol S is proposed for listing on California Proposition 65. In this context, the authors attempted to study on different phenolic monomers present in aromatic syntans and their effects on leathers. The effect of monomers on the haptics, light fastness, heat yellowing, and residual floats was studied. The study has discovered the small increase at tightness and hardness in leathers with the increase of monomers phenolsulfonic acid and bisphenol S at low quantities ~2 wt%, while at >10 wt% monomers in aromatic syntans the leathers have become emptier. Further, the research has revealed that monomers have the following order of influence on the light fastness: Phenol>bisphenol S =bisphenol F>phenolsulfonic acid. E.g., Leathers retanned with an aromatic syntan that was containing a free phenol concentration of <50 ppm has a Gray Scale (GS) value of 4.0 which is reduced to a GS 2.0 when a syntan containing >4,000 ppm of free phenol was applied. The study also showed that there was no significant effect of phenolic monomers on the heat yellowing of the leathers. Bisphenol S and F showed an uptake of above 90% while phenol and phenolsulfonic acid have showed values <40%, where the remainder was flushed into the waste waters.

The research study consist of 3 different parts:

(i) Amount of monomers: This part discusses the amount of phenolic monomers in different aromatic syntans across the market.

(ii) Model study: A research study has been conducted to determine the effect of each specific monomer in a specific syntan on the leather. In this study, several model syntan samples were prepared by addition of different loadings of each monomer, in order to determine the effect of each monomer separately.

(iii) Effects of monomers: In this part an attempt has been made to correlate the combined effects in different syntans across the market to the amount of monomers present in them.

Keywords: (re)tanning, aromatic syntan, waste waters, phenolic monomers, bisphenol S, bisphenol F, phenol, phenolsulfonic acid and restricted compounds

1. Introduction

Aromatic syntans play an important role in the retanning step or wet end process of leather making. Syntans modify the haptic properties of the leather and give the required softness and fullness to the leathers during the retanning. Syntans are widely used for all types of tanned leathers, namely Chrome (Wet Blue, WB), Glutaraldehyde (Wet White, WW) and vegetable tanned leathers to give the additional strength [Ammenn. J et al. 2015, Murali. S et al. 2019, and Song. J et al. 2018]. The amount of syntan used varies upon the process (retanning or sole tanning), condition of hide/skin, and type of article that will be produced. Aromatic syntans, being synthetic in nature, are well suited for all color shades and result in leathers with a better lightfastness, compared to leathers retanned with vegetable tannins.

In general aromatic syntans are condensates of phenolsulfonic acid and/or phenol and/or dihydroxydiphenylsulfone (DHDPS or bisphenol S) with formaldehyde where urea is optionally used. Depending upon the efficiency of the condensation process, a certain amount of phenolic monomers are left unreacted in the production process. These unreacted phenolic monomers such as phenol, phenolsulfonic acid and bisphenol S/F that are present in the syntans, will be transferred to the leather during the (re)-tanning processes. The presence of unreacted monomers are a problem with respect to the following three main aspects. They are:

(i) Hazard classification of the each monomer and product regulatory aspects

(ii) Efficiency during the (re)-tanning of the leather (monomers either ending up in effluent or causing an undesired effect on the leather) and,

(iii) Effect on the leather haptics and leather properties.

The current article has tried to investigate the important aspects around the amount of phenolic monomers. Namely:

1. What is the amount of monomers in the aromatic syntans across the leather industry?

2. What is the effect of monomers on the leather haptics and leather properties?

3. What are the current regulations stating on these monomers and the impact on the leather industry

Identification and determination of the phenolic monomers in the syntans is the first step for future optimizations, and to produce cleaner products which have less impact on heath safety and environment (HSE) aspects.

2. Materials and methods

2.1 Chemical analyses

2.1.1. HPLC (High performance Liquid Chromatography) analyses to determine the phenol and phenolsulfonic acid in retanning agents

Methanol of HPLC grade was purchased from Chem-Lab; while phenol, phenolsulfonic acid and tetrabutylammonium bisulfate (analytical grade) were purchased from Merck. Depending on the expected concentration of analytes, 0,5-5 gram of sample was dissolved in 20 ml Milli Q water using a magnetic stirrer. This solution was quantitatively transferred to a 50 ml volumetric flask and filled. An aliquot of this solution was filtered through a 0.45 µm filter (Millipore Phenex RC) and transferred to a 2 ml HPLC vial. The HPLC analyses was performed at a Shimadzu LC-20AD Prominence equipped with a UV-vis photodiode array detector (SPD-M20A), an on-line Degasser filter (DGU-20A5), and a 20-µL sample loop. A Phenomenex Gemini NX C18 (4,6 mm x 250 mm x 5 µm) was used. 10 µl of the sample solutions were injected into the system, using a flowrate of 1,0 ml/min at a column temperature of 40°C, using an isocratic eluent composition of 15% methanol and 85% 0.03M TBAS in Milli Q water for 30 minutes. Quantification of the phenol and phenolsulfonic acid peaks were performed by using external calibration of the peaks at 13,5 (phenolsulfonic acid) and 23,0 (phenol) minutes.

2.1.2. Determination of bisphenol F and S in retanning agents and leathers

Syntan samples and leather samples are outsourced for the analyses of bisphenol F/S at FILK Freiberg institute, CTC Lyon and Ars Tinctoria SRL Santa Croce. The used methods for the analyses are in-house developed methods at the respective institutes.

2.2 Leather properties analyses

Leather properties recipes: The recipes to determine the leather properties (heat yellowing and lightfastness) are different compared to typical wet end recipes, which are also carried out as per Smit & Zoon internal testing methods. These single recipes are executed without dye and fat liquor, but with syntan and are designed to exclusively determine the effect of the syntan

2.2.1 Determination of the heat yellowing of the leathers

Conditioning (heating) of leather pieces (6,0 x 7,0 cm) in a ventilated oven for various time periods such as 24, 48, 72, 144 h and at 100 and 110 °C temperatures. Color determination according to ISO 105-A05:1996, performed using a Xrite SP60 portable spectrophotometer.

2.2.2 Determination of the lightfastness of the leathers

Radiation of leather pieces (6,0 x 7,0 cm) using an Atlas Suntest CPS+ at 50°C for 72 hours. Color determination according to ISO 105-A05:1996 was performed using a Xrite SP60 portable spectrophotometer.

2.3 Haptic properties

Recipes: All the leather tests are carried out as per internal Smit & Zoon wet end recipes. Leather applications for haptic performances are typically used standard recipes where dye, fat liquor and other auxiliary leather chemicals are used along with syntan powder or liquid.

Leather assessments: Each haptic property like softness, fullness, grain tightness etc. is assessed on a 1 to 7 scale. Experimental leathers are within this procedure always assessed with a comparative leather as a reference. Leathers assessments are performed by a minimum of 4 technicians and the ratings are averaged to present the final assessment of the leathers for each haptic property separately.

2.4. Analyses of Waste waters/Uptake

Waste waters are analyzed for phenol and phenolsulfonic acid by HPLC as described in the chemical analyses section 2.1. Waste water analyses are used to indirectly estimate the uptake of phenol and phenolsulfonic acid on to the leathers. Alternatively, the uptake of bisphenol S and F is measured by direct analysis of the leathers, where the bisphenol S and F were first extracted from the leathers, before being analyzed by HPLC.

3. Results

3.1. Amount of phenolic monomers in different aromatic syntans

Several aromatic syntans from across the market are analyzed to determine the small amounts of phenolic monomers. HPLC was used to determine the amount of free monomers, more details can be found in the materials and methods section. Table 1 presents the amounts of phenolic monomers in sulfone based syntans (sulfone condensates, SC) series. The amount of monomers in liquid syntans are corrected to their solid content in order to make a comparison with the powdered syntans.

All sulfone based syntans have bisphenol S amounts above 1 wt% (10,000 ppm), and sometimes as high as 6 wt%. When it comes to free phenol, most sulfone syntans have a concentration lower than 100 ppm (see Table 1). Similarly, bisphenol F concentrations in sulfone syntans are usually also very low, with typical values around 100 ppm for the most sulfone based syntans, barring few exceptions where the amount ranges up to 600 ppm. The amount of phenolsulfonic acid varies from one syntan to another. Two sulfone syntans with a relatively high bisphenol S monomer concentration, contain very low phenolsulfonic acid amounts (<0,1 wt%), while most others have around 1 to 2 wt% phenolsulfonic acid with the two exceptions and which have around 4 and 13 wt%. Relative high amounts of bisphenol S can be attributed to the process chemistry of sulfone syntans.

Table 2 presents the amount of phenolic monomers in phenol condensates (non-sulfone condensates) which are typically phenolsulfonic acid and/or phenol condensates. Due to their difference in chemistry compared to sulfone based syntans, phenol condensates have very low bisphenol S amounts, which are in the range of 100-200 ppm. Alternatively the phenol condensates contain free phenol concentrations in the range of <20 to 8,000 ppm and bisphenol F amounts ranging from 100 to 12,000 ppm. Remarkably, phenol condensates exist with both very low bisphenol F and phenol concentrations. The amount of phenol and bisphenol F monomers in phenol condensates seem to have a proportional relationship. When the phenol monomer concentration is relatively low, automatically bisphenol F concentration is also low (Figure 1).

3.2. Uptake of monomers in the leathers

Bisphenols S and F: Selected leathers made by leather properties recipes are tested for the amount of monomers. An extraction method followed by HPLC measurement was used to determine the monomers in the leathers and leathers produced with SC-4L and PC-2L are tested for the presence of monomers. The uptake of bisphenol S/F are both determined to be above 90%, when compared to the corresponding maximum amount based on the total of bisphenol S/F in the syntans. These results show a very high uptake for these monomers.

Phenol and phenolsulfonic acid: wastewaters from the full recipes of syntans with spiked phenols are analyzed for free phenols. Uptake of phenols into the leathers is indirectly calculated based on the measurement of free phenol in the waste waters. The results show that the uptake of phenol was approximately 30 to 40 wt% of the corresponding maximum amount of free phenol present in the syntan. Like phenol, the uptake of phenolsulfonic acid was calculated from the phenolsulfonic acid that was found in the waste waters. Accordingly phenolsulfonic acid's uptake was calculated as 20-30% to the corresponding phenolsulfonic acid present in the syntan.

3.3. Effect of monomers on the haptics of the leather

Syntan SC-4 L was spiked with different amounts of phenol to check the effect of the syntans on the leather. The resultant syntan blends are tested on WW leathers. The application of the newly produced syntan (SC-4+ with 10,000 ppm phenol) produced somewhat softer, somewhat fuller and somewhat lighter (in color) leathers compared to SC-4 (<20 ppm free phenol). Small amounts of spiked free phenol <1,000 ppm in SC-4 have shown minimal changes in leathers compared to the reference. Results indicate that free phenol is helping in the retanning process of the leather.

Syntan SC-12 was spiked with different amounts of bisphenol S monomer. The resultant modified syntans SC-12+ with 3,6 wt% bisphenol S and SC-12 with 6,7 wt% bisphenol S are tested as retanning agents against SC-12 with 2,4 wt% bisphenol S on WW leathers. 3,6 wt% and 6,7 wt% bisphenol S containing SC-12 have produced somewhat lighter (color intensity), somewhat harder, somewhat tighter and somewhat fuller leathers.

PC-6 was used to test the effect of the influence of free phenolsulfonic acid on the leather. PC-6 (with 2,2 wt% phenolsulfonic acid) was spiked with different amounts of phenolsulfonic acid to make 4,0 wt%, 5,2 wt% and 8,9 wt% phenolsulfonic acid containing syntan blends. With the increase in the amount of phenolsulfonic acid in the PC-6, we see the increasing effects at hardness, emptier handle, tighter grain, and uneven milling (chicken wires) on the leathers. The monomers bisphenol S and phenolsulfonic acid are assumed to not fully penetrate, but to bond to the surface of the leathers and thereby causing additional hardness. When the monomer concentration of both bisphenol S and phenolsulfonic acid was even more increased, the produced leathers also started showing a decrease in fullness. This behavior is assumed not to be caused by the presence of the monomers, but indirectly by the decrease of the concentration of the polymeric species, which are the main active species for the retanning.

3.4. Influence of monomers on the lightfastness and HF of the leathers

3.4.1. Influence of bisphenol S monomer in syntans:

Figure 2 displays the lightfastness (GS values) of WW leathers prepared by sulfone based syntans containing various amounts of bisphenol S. Results revealed that there is an inversible proportional relationship to the amount of bisphenol S in the syntan to the lightfastness. To confirm this effect, different amounts of bisphenol S monomer was spiked to a specific syntan (SC-12) and the resultant syntan blends are tested for their lightfastness. Contradictory to our expectation, lightfastness of the sulfone syntan blends with varying amounts of spiked bisphenol S, have resulted to produce minimal changes to the lightfastness of the leathers (see the Figure 4).

Varying bisphenol S amounts in the syntans, have not shown any trends on towards the heat yellowing resistance (no figure is presented). Based on this can be concluded that free bisphenol S has minimal influence on the heat yellowing in the leathers.

3.4.2. Influence of phenol in syntans

Figure 3 shows the lightfastness (GS values) of WB leathers prepared by various phenol condensates containing different amounts of free phenol. Results have showed an inversible proportional relation of the free phenol concentration in the syntan to the lightfastness of the treated leathers. To confirm this effect, different amounts of phenol monomer was spiked to PC-3 L and the resulting PC-3 L+ treated leathers were tested for lightfastness. Like in the case of bisphenol S, here too, different amounts of spiked phenol containing syntans have shown minimal changes in the lightfastness of the leathers (Figure 5).

3.3.3. Influence of Phenolsulfonic acid in syntans

Figure 6 presents the effect of the spiked phenolsulfonic acid in the PC-6. Results indicate that free phenolsulfonic acid also has a minimal effect on the lightfastness of the leathers.

4. Discussion

4.1. Efficiency during the (Re)- tanning on the leather:

Many aromatic syntans have total monomer concentrations in the range of 2-5 wt%, while very few are in the range of 10 wt% or above. Based on the current studies, as well from earlier studies, it is clear that polymers are the ones that have more (re)tanning effect [Ammenn. J 2009] compared to the monomers. Monomers are effecting leather haptics and waste waters negatively. Leather chemical producers need to constantly work towards optimization of their processes and to minimize the amount of leftover monomers and the resultant products have less wastage and show in turn more efficiency on the leathers.

4.2. Impact of phenolic monomers (in the aromatic syntans) in leather chemicals:

From the four phenolic monomers researched, phenolsulfonic acid shows to be the least problematic with regard to possible health issues and classification. According to the European Chemicals Agency (ECHA) database, phenolsulfonic acid is labelled as corrosive, and the acute toxicity of the chemical is quite low (LD50 = 1410 mg/kg).

According to the Hazard classification & labeling by ECHA, phenol shows high acute toxicity (LD50 = 340 mg/kg) and suspected to be mutagenic, which is of high concern when high concentrations are present in retanning agents. Phenol is widely researched for its adverse effect on health [Gami. A. A et al., 2014; Yonghong. X et al. 2006]. There are no limits on the amount of phenol for retanning agents now. However, phenol is in the ZDHC MRSL 2.0 candidate list for textiles, which will in the future result in a maximum allowed concentration of this substance in the produced chemicals. A similar future restriction level can be expected for leather chemicals. Several consumer brands have been asking on the amount of phenols in the retanning agents and leathers.

Bisphenol S has no ECHA hazard classification & labeling now. Bisphenol S shows very low acute toxicity (LD50 = 2830 mg/kg) and suspected to be toxic to the reproduction system and under assessment as endocrine disrupting. This is also evident from several recent research studies [Fouyet. S et al. 2021]. Research studies have demonstrated that bisphenol S shows very similar toxicity behaviour as bisphenol A, therefore a strong argument may be made to regulate bisphenol S in exactly the same manner as BPA. There is a regulation proposal in front of ECHA for all the bisphenols, which also includes bisphenol F.

Currently there is little impact on the industry in relationship to bisphenol S. However there is an intention for a harmonized classification and labelling of bisphenol S as reprotoxic 1B by European Union. This would mean that all products containing >0,3% bisphenol S will be classified and labelled accordingly. This classification would also qualify bisphenol S as substances of very high concern (SVHC). Once the restriction is published, a transition period will be included in the legal text. Once the proposed restriction comes into force, (leather) articles containing >0.02% of any bisphenol are prohibited in the EU. This limit will be for individual bisphenols and there is no indication that there will be a summation limit for all bisphenols combined. In a future timeline of 5 to 7 years, these restriction can come into place. Currently, consumer brands are already asking for a reduction of bisphenol S concentrations in leathers.

4.3. What are the compounds in the syntans that are effecting the lightfastness?

Lightfastness of the leathers prepared using syntans, is also an important subject. Many different compounds in syntans can have influence on the lightfastness of the leathers. Among all, some compounds can have bigger influence than others. Based on the initial results of this research, it is believed that the free phenol and bisphenol S in the syntans have inversely proportional relationship to the lightfastness of the syntans. However it is evident from the later studies that neither phenol nor bisphenol S have direct effect on the lightfastness of the syntans. Bisphenol S and phenol are believed to be marking compounds for the inversely proportional effect on the lightfastness of the leathers. Real compounds that are effecting lightfastness are currently hypothesized as intermediates during condensation(2 or 4-hyroxymethyl phenol and 2-hydroxymethyl bisphenol S), which are formed during the condensation reaction; The concentration of these compounds in syntans, is expected to be in the same range compared to phenol and bisphenol S. Hence the concentration of phenol and bisphenol S could have shown inversely proportional to the lightfastness. A previous work on the lightfastness of the syntans has demonstrated the positive effect of the urea in the condensation along with phenol and formaldehyde; where phenol, urea and formaldehyde condensed products show to provide a better lightfastness in the leathers compared to phenol and formaldehyde leathers [Venkataboopathy. K et al. 2001]. These results further confirm the current hypothesis on the effect of compounds contributing to the poor lightfastness.

4.4. Challenges to sulfone based syntans:

Sulfone based syntans contain relatively high amount of bisphenol S in the syntans. This is correlated to the relatively low reactivity of bisphenol S in comparison to other phenolic monomers, such as phenol. Additionally, bisphenol S has shown to have a very high uptake on the leather and both factors combined make it quite challenging to produce sulfone based syntans that, after application in the leather production recipe, result in leathers with a bisphenol S concentration of less than 200 ppm. Chemical producers of sulfone based syntans have to either minimize the amount of bisphenol S by optimization of the current process or develop a new process to produce similar products without free bisphenol S. There is already research in this direction to minimize the bisphenol S [Zhang. S et al. 2021]

4.4. Future of phenol condensates:

When comes to the phenol condensates, phenol shows to be relatively reactive. By utilizing this advantage, the producers of phenol condensates have a good possibility to optimize the production processes towards an extremely low concentration of free phenol in the syntans. Phenol has a moderate uptake into the leather, but this also means that a big proportion of the free phenol present in the syntans will end up in the waste waters. When it is possible to reduce the phenol concentrations in syntans sufficiently, there is a strong reason to prefer phenol condensates over sulfone condensates.

5. Conclusions

Sulfone based syntans across the market have relatively high amounts of bisphenol S (1 to 6 wt%), while they contain only small amounts of phenol (<100 ppm).

Phenol condensates (non-sulfone based syntans) have free phenol concentrations in a wide range (<20 to 8000 ppm) and bisphenol F in proportional amounts.

Amounts of bisphenol S and phenol in the syntans are appeared as inversely proportional to the lightfastness of the leathers. However model syntans have suggested these are only markers while the real responsible compounds for the poor lightfastness are different and which are hypothesized as intermediates during condensation(like 2- or 4-hyroxymethyl phenol and 2-hydroxymethyl bisphenol).

Leathers have shown to become harder and tighter, with the increase of the amount of phenolsulfonic acid and bisphenol S in the syntans. In the syntans where monomer concentrations are increased to above 5 wt%, the produced leathers have shown to become emptier. Conversely, with the increase of phenol in the syntans, leathers show to become softer and fuller.

Results have demonstrated that phenol condensates with lower concentrations of free phenol, are able to create leathers with higher lightfastness properties in comparison to sulfone based syntans.

Bisphenol S and bisphenol F both have an excellent uptake into the leather, which is determined to be above 90% corresponding to the maximum available amount of monomers in the syntans. Phenol shows a moderate uptake into the leather, which is of around 30-40 wt%.

Based on the toxicity studies, a restriction proposal on all bisphenols is under assessment at ECHA. There is a strong indication that bisphenol S and F, will also get the similar indication as bisphenol A, that is labeled as both reprotoxic and endocrine disrupting.

The focus of the syntan market could be shifted from sulfone condensates towards phenol condensates when manufacturers of the phenol condensates are able to optimize their production processes to create low phenol containing syntans.

6. The acknowledgements

Authors are grateful to the colleagues who contributed to the research of this paper. Arnaud Backbier, Job der Kinderen, Martin Heise, Steffen Kohl and Lars Torpus from Product Management for the help at the organization of the analyses for big number of samples; Ad Elshout, Sarah Nadi and Tomasso Trabellesi from Leather application center for the execution of different leather applications; Gerjanne Tiggelaven, Rob Meulenbroek and Teenu Tomy from the analyses group for several in-house analysis. Next authors are thankful to the Lilian Chang, Global R&D Director of Smit&Zoon for the excellent support and encouragement throughout this research and allowing us to present this research work at 5th WLC and XXXVI IULTCS Congress Addis Ababa Ethiopia.

List of references

Ammenn. J, Glocknitze. F, Tiarks. F, Wolf. G, (2009) Structure-Activity-Relationships of Synthetic Organic Tanning Agents., Oral presentation 1-28, Proceedings of the XXX IULTCS congress Beijing, China

Ammenn. J, Huebsch. C, Shilling. E, and Dannheim B (2015): Chemistry of Syntans and Their Influence on Leather Quality. Journal of American Leather Chemicals Association., (110)349

Fouyet. S, Olivier. E, Leproux. P, Dutot. M, Rat. P (2021): Bisphenol A, Bisphenol F, and Bisphenol S: The Bad and the Ugly. Where Is the Good?. The Life.,(11) 314

Gami. A. A, Shukor. M. Y, Khalil. K.A, Dahalan. F. A, Khalid. A, Ahmad. S. S (2014) Phenol and its toxicity, Journal of Environmental microbiology and toxicology 1, 11

Murali. S, Subramanian. B, R. R. Jonnalagadda, N. N. Fathima (2019): Deciphering the Role of Individual Retanning Agents on Physical Properties of Leathers. Journal of American Leather Chemicals Association., (114) 94

Song. Y, Wang. Y.-N, Zeng. Y, Wu. H, Shi. B. (2018) Quantitative determinations of Isoelectric point of retanned leather and distribution of retanning agent Journal of American Leather Chemicals Association.,113, 232

Venkataboopathy. K, Prabhu. R, Meenakshi. S and Sundar Rao. V.S. (2001), Preparation and Chracterization of Synthetic Tannins based on Phenol-Dimethylolurea resin., Journal of the Society of Leather Technologists and Chemists 85, 175

Yonghong. X, Xinshen. Z, Wenxian. W, Jin. H and Liang. W. (2006).,, Determination of volatile phenol in Tannery effluent., Journal of the Society of Leather Technologist and Chemists., 92, 30

Zhang. S, Jiang. J.-Q, and Petri, M. (2021), Preliminary comparative performance of removing bisphenol-S by frerrate oxidation and ozone npj clean water 4:1

Table 1: Amount of phenolic monomers in different sulfone based syntans across the Leather industry*

Type of Syntan	Phenol, in ppm	Phenolsulfonic acid, in wt%	Bisphenol F, in ppm	Bisphenol S, in wt%
SC-0	<20	1,5	300	0 (<75 ppm)
SC-1	<100	0,5	600	2,6
SC-2 L	<100	13,6	ND	4,3
SC-3	100	1,9	ND	2,6
SC-4 L	<20	3,3	<75	2,2
SC-5	2000	2,6	ND	1,9
SC-6 L	<20	0,1	<75	6,1
SC-7	<100	2,2	<75	1,5
SC-8 L	220	1,4	ND	2,4
SC-9	90	0,9	ND	1,6
SC-10 L	50	4,4	<20	1,5
SC-11 L	<20	0,1	ND	6,4
SC-12	<50	2,1	<75	2,4

*HPLC method was sued to determine the monomers in the syntans

*The monomers weights in the liquids are corrected to dry weight

ND= not determined

Table 2: Amount of phenolic monomers (in ppm) in different phenol based syntans across the Leather industry*

Type of Syntan	Phenol, in ppm	Phenol sulfonic acid, in wt%	Bisphenol F, in ppm	Bisphenol S, in ppm
PC-1	1500	8,6	2000	<75
PC-2 L	100	2,5	1500	<75
PC-3 L	<20	1,5	300	<75
PC-4	5000	4,7	12000	<75
PC-5	3000	ND	10000	100
PC-6 L	500	3,0	1100	<50
PC-7	<20	0,0	2000	ND
PC-8	<20	0,03	1800	ND
PC-9	200	0	590	0
PC-10	4500	3	10000	0
PC-11	7600	1,4	10500	0

*HPLC method was sued to determine the monomers in the syntans

*The monomers weights in the liquids are corrected to dry weight

ND= not determined

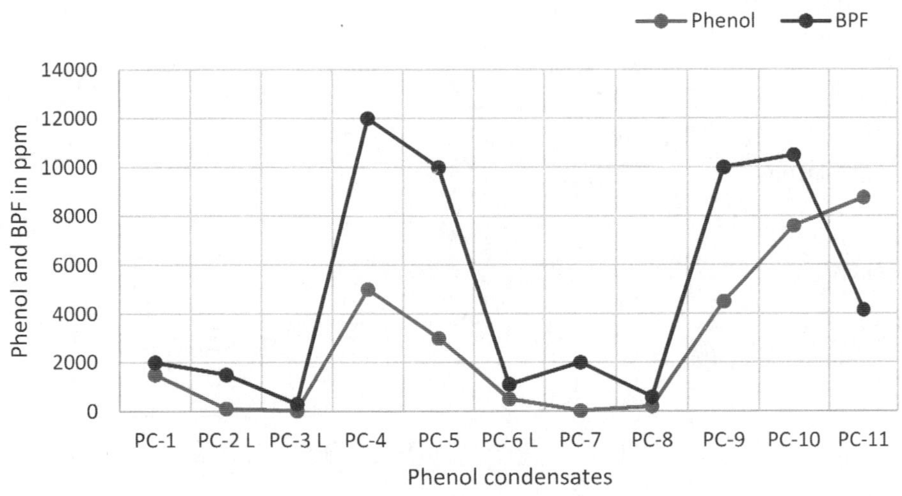

Figure 1: Amounts of phenol and bisphenol F (ppm) in the phenol condensates

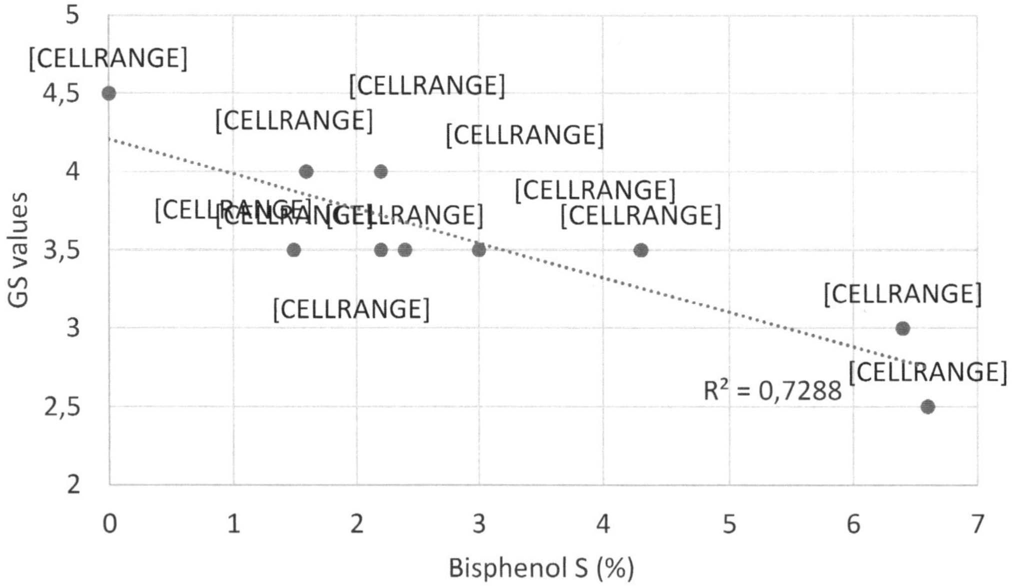

Figure 2: Lightfastness of WW leathers prepared by various sulfone based syntans. Figure presented the amount of free bisphenol S (wt%) versus Grey Scale (GS) values collected at 72 h/BST 50°C (Blank WW leather GS value =3,5)

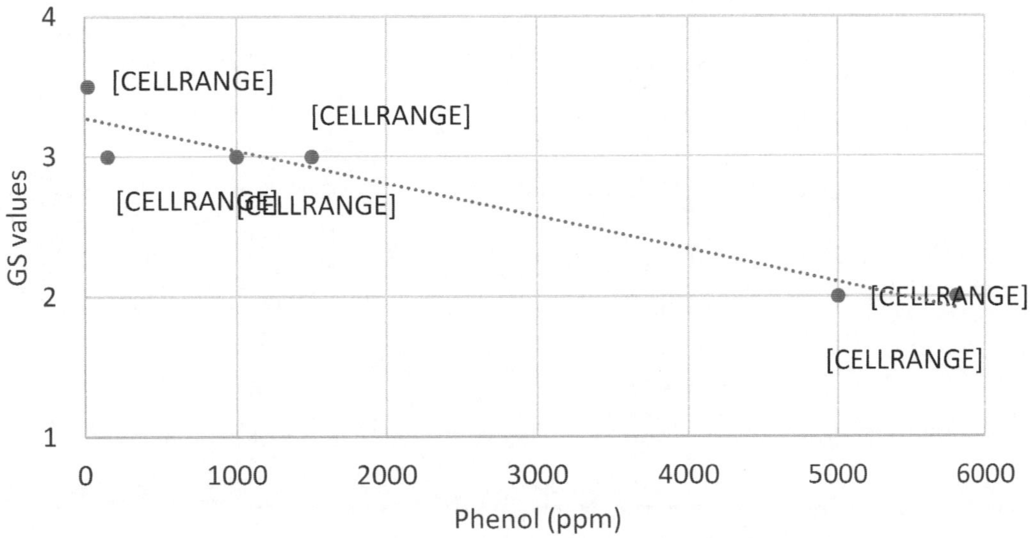

Figure 3: Lightfastness of WB leathers prepared by various phenol condensates. Figure presented the amount of free phenol (in ppm) versus Grey Scale (GS) values collected at 72 h/BST 50 °C (Blank WB leather GS value =4,5)

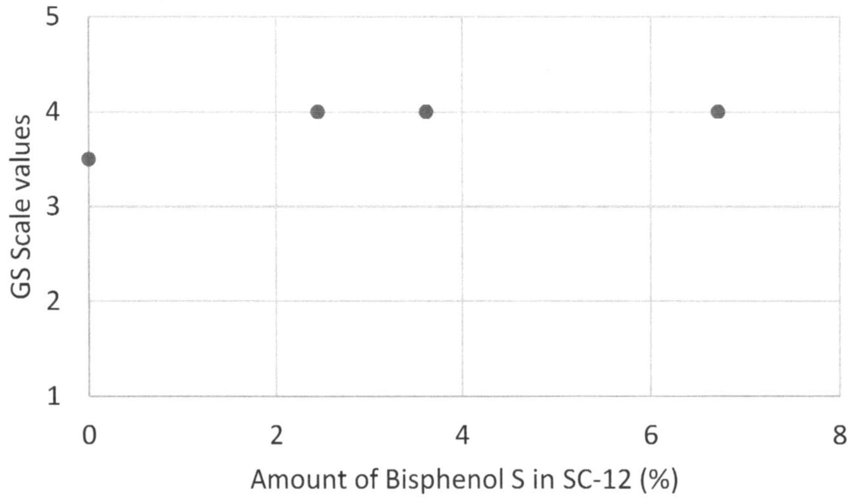

Figure 4: Lightfastness of WW leathers prepared by the syntan blends prepared by SC-12 by spiking different amounts of Bisphenol S. Graph depicted the amount of Bisphenol, wt% versus Grey Scale values collected at 72 h/BST 50 °C (Blank WW leather GS value = 4,0)

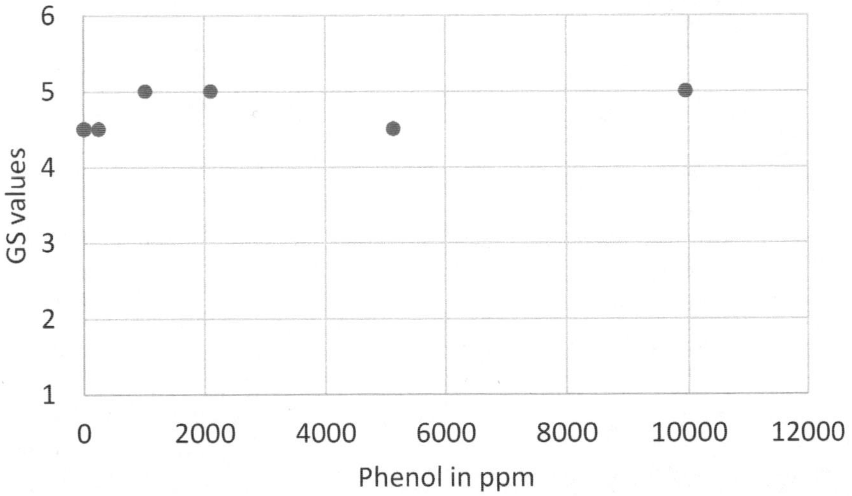

Figure 5: Lightfastness of WW leathers prepared by the syntan blends prepared by PC-3 L (free phenol <20 ppm) by spiking different amounts of free phenol. Graph depicted the amount of free phenol (in ppm) versus Grey Scale values collected at 72 h/BST 50 °C (Blank WB leather GS value =4,5)

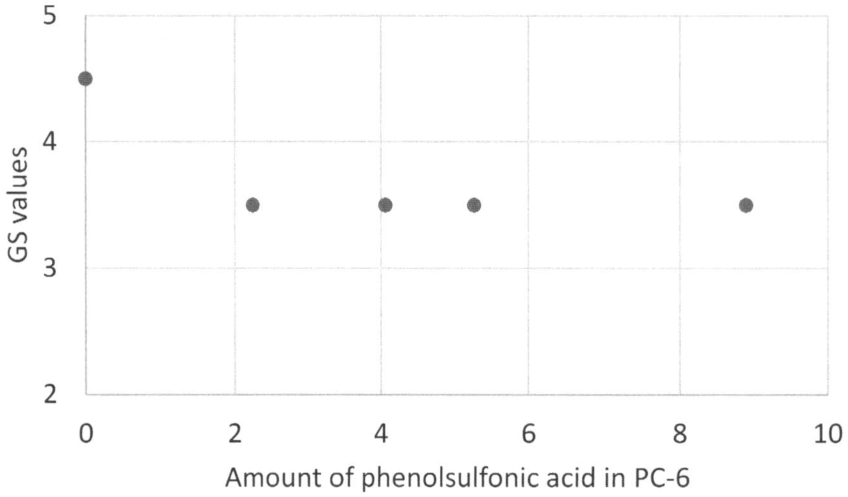

Figure 6: Lightfastness of WB leathers prepared by the syntan blends prepared by PC-6 by spiking different amounts of phenolsulfonic acid. Graph depicted the amount of phenolsulfonic acid, wt% versus Grey Scale values collected at 72 h/BST 50 °C (Blank WB leather GS value = 4,5)

A novel preparation of formaldehyde free melamine resin and its retanning behaviors investigation

Ji-bo Zhou [a], Guo-qiang Ning [a], Xue-pin Liao [a,b※] and Bi Shi [a,b]

[a] The Key Laboratory of Leather Chemistry and Engineering of Ministry of Education, Sichuan University, Chengdu 610065, PR China

[b] National Engineering Centre for Clean Technology of Leather Manufacture, Sichuan University, Chengdu 610065, PR China

Melamine resin (MR) retanning agent was widely used in leather industry, but often suffered the problem of formaldehyde release. Although numerous efforts have been devoted to solving the formaldehyde issue by using alternative aldehyde compounds, it often encounters a new challenge of poor storage ability due to their relative low reaction activity compared with formaldehyde. Herein, a new formaldehyde-free melamine resin MGE was prepared through a novel synthesis strategy. Melamine was firstly bonded with lysine to improve its amphoteric and water solubility, and the epoxy compound was further used as a cross-linker to prepare melamine resin. Comprehensive

* Corresponding author. National Engineering Centre for Clean Technology of Leather Manufacture, Sichuan University, Chengdu, 610065, China. e-mail address: xpliao@scu.edu.cn

structural and performance characterization of MGE confirmed the presence of amphoteric groups in its structure. Its isoelectric point was determined to be 7.55, significantly higher than the other two commercial MR which are 2.94 and 4.69, respectively. Moreover, MGE showed high storage stability of more than six months. The physical and mechanical properties of MGE retanned leather, along with its reactivity to anionic fatliquor were also fully investigated. Owing to the ionic bonds and hydrogen bonds between MGE and collagen fibers, MGE-retanned leather showed improved physical and mechanical properties. The absorption of MGE was up to 77.08% and the thickening rate was increased by 19.23%. This study provided a new strategy to synthesis formaldehyde-free melamine resin with high stability and excellent retanning performances.

Keywords: leather, retanning agent, melamine resin, formaldehyde free, synthesis

1.Introduction

Melamine-formaldehyde resin (MR) has been widely used in leather industry for its excellent filling ability. However, the formaldehyde issue has severely hindered its further application, because more and more stringent regulations have been established to limit formaldehyde content in leather products. Therefore, efficient alternatives are highly needed to solve the formaldehyde problem of MR(Arias, González-García, Feijoo, & Moreira, 2021; Kord, Movahedi, Adlnasab, & Ayrilmis, 2021).

Extensive works have been conducted in the past few years to address the formaldehyde problem in MR. Approaches explored by researchers can be divided into two categories(Akinyemi, Olamide, & Oluwasogo, 2019). The first one involves changing parameters in the synthesis of MR, such routes including adjusting the percentage of reactants, introducing other amino compounds, and changing the procedures of reactions in batch(J. Li & Zhang, 2021; Song et al., 2021). These methods can only reduce but not completely eliminate the formaldehyde. The second method involves using environmentally friendly reactants in the synthesis of formaldehyde-free resins, the starting compounds including glyoxal, furfural, and some natural feedstocks such as starch, tannin, and protein-based adhesives(Ji-bo, Jian-fei, Pei-lin, Xue-pin, & Bi, 2018; Kan, Sun, Kan, Bai, & Gao, 2021; Okumura, 2021; J. Wang, 2021). These synthesized MR have no aforementioned health issues, but they are mostly in the laboratory development stage and still suffer from several obstacles such as lower reaction activity, much higher cost, or complicated synthesis procedures(Balducci et al., 2020). There still remains a great challenge to find a pragmatic pathway to realize the zero emission of formaldehyde in MR.

Glutaraldehyde is a well-known crosslinker with highly reactivity, and it is suitable to substitute formaldehyde in some regards. Regrettably, the reaction between glutaraldehyde with melamine must be sulfonation first to improve the water solubility of MR. But this process may decrease the positive charge of MR and finally influence the absorption extent of the following anionic fatliquoring agents and dyestuffs(Liu, Wang, Wang, Sun, & Wei, 2021). Therefore, the effective application of glutaraldehyde in the preparation of MR is still a challenge.

Lysine is a naturally available renewable feedstock and contains a carboxyl group and two amino groups in its molecule. The amino groups make lysine very easy to bond with melamine by glutaraldehyde. By this approach, the lysine can not only improve the water solubility but also confer amphoteric property to final MR products. Moreover, lysine supplies good compatibility with collagen fibers, thus may result in a higher absorption of synthesized retanning agents. In light of the above, it is reasonable to believe that the introduction of lysine may confer good performances to MR.

In work described here, a new formaldehyde-free melamine resin MGE was prepared through a novel synthesis strategy. According, melamine was firstly bonded with lysine by glutaraldehyde. Then, ethylene glycol diglycidyl ether (EGDE) was further used as a cross-linker in the synthesis of formaldehyde-free MR. EGDE is one of the most important commercially compounds, it has versatile reactivity and can be severed as a substitute for the polycondensation stage to realize the crossing-linking process(Cai et al., 2021). The chemical structure of prepared MGE was characterized by FT-IR and ^{13}C-NMR, and its application properties in leather retanning process were fully investigated.

2. Materials and methods

2.1 Materials and equipments

Melamine and glutaraldehyde were analytical grade and purchased from Chengdu Kelong Chemical Co., Ltd (Chengdu, China). Lysine and ethylene glycol diglycidyl ether (EGDE) were analytical grade and purchased from Guangdong Wengjiang Chemical Reagent Co., Ltd (Shaoguan, China). Wet blue leather (cow hide) shaved to a thickness of 1.0mm was obtained from a local tannery.

2.2 Preparation of MGE

Aqueous solution of glutaraldehyde (50% by weight) was used in this study and MGE was prepared as following steps. First, pH value of the glutaraldehyde solution was maintained to 8.0–8.5 by using 1 M NaOH solution and taken into a 250 mL three-necked flask equipped with magnetic stirrer. The reaction bath was kept at 60°C. A definite mole proportion of melamine was then added to the flask and stirred until melamine is completely dissolved. This process was recognized by the color change from a white suspension to a clear solution and the intermediate products was named MG. After that, lysine was dissolved by an equal mass of water and added to the system. The reaction was continued for further 30 min after the addition of lysine and the intermediate product of this procedure was named MGY. Then, the pH of MGY was adjusted to 5.0 by using 1 M HCl solution and different rations of EDGE (1%, 2.5% and 5% based on total weight of MGY, respectively) were added dropwise with vigorous stirring and kept for 120 min, the products were named MGE1, MGE2 and MGE3, respectively. Finally, the clear solution was cooled down naturally and MGE was obtained. The synthetic scheme was illustrated in Figure 1.

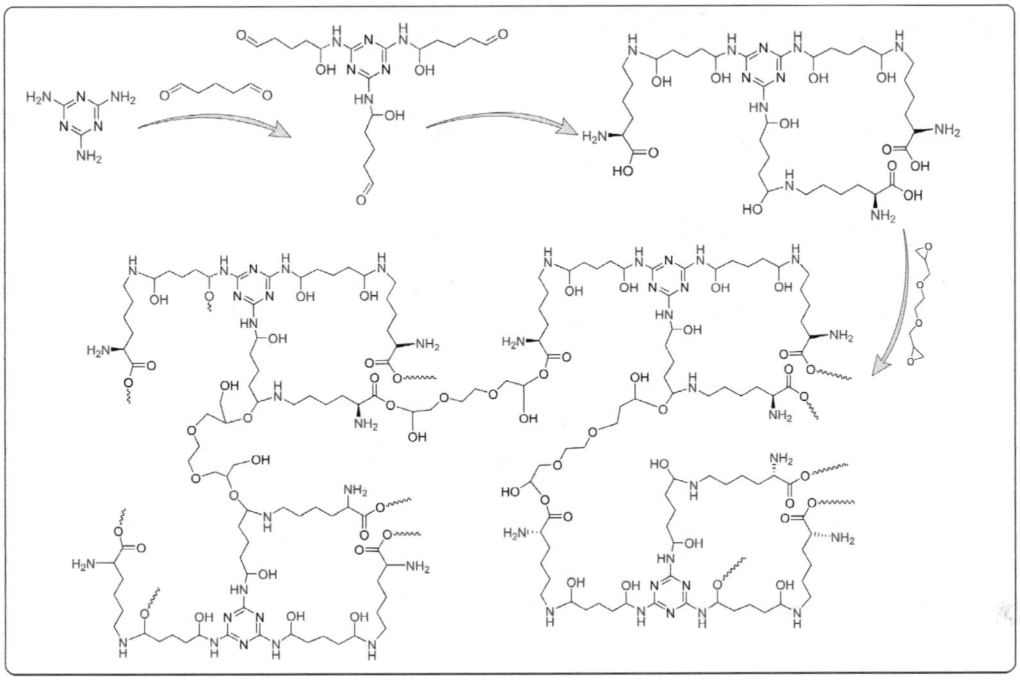

Figure 1. The schematic representation of MGE preparation processes

2.3 Characterization

2.3.1 FT-IR Analysis

The FT-IR spectra of melamine, MG, MGY and MGE were performed in a Nicolet IS10 (Thermo Scientific, USA) spectrometer. The test was carried out using compressed pellets of KBr and the wavenumber range from 400 to 4000 cm^{-1}.

2.3.2 Nuclear magnetic resonance spectrum (^{13}C-NMR)

Nuclear magnetic resonance (NMR) spectra were recorded on a Bruker Avance II 400MHz NMR spectrometer (Bruker, Swiss). The solvent was heavy water (D_2O).

2.3.3 Particle size and isoelectric point (pI) analysis.

The average particle size and the zeta potential measurements were determined by NanoBrook Omni (Brookhaven, USA). Samples were dissolved in water to a final concentration of 1g/L and their pH were adjusted 2-10 with 1mol/L HCl and 1mol/L NaOH to further determine the isoelectric point (pI). Two commercial amino resin retanning agents named CMS and CML were used as control samples.

2.4 Application performances test

For this study, the wet-blue (chrome tanned) was used as raw materials to evaluate the application performances of prepared MGE. The leather samples were cut into two halves through the backbone and divided into 30 × 30 cm, their thicknesses were about 1.2 mm. Moreover, two commercial amino resin retanning agents named CMS and CML were used as control. The application processes were summarized in Table 1.

Table 1 Retanning Processes

Process	Chemicals	%	Duration(min)	Remarks
Rewetting	Water	400		
	Degreasing agent	0.3	60	Drained
Washing	Water	400	10	Drained
Neutralization	Water	200		
	Sodium formate	2	30	
	Sodium bicarbonate	0.4	10	
	Sodium bicarbonate	0.4	60	
Washing	Water	400	10	Drained
Retanning	Water	100		
	CMS/CML/MGE	8	60	
Fatliquoring	Synthetic fatliquor	10	30	
	Hot water	100	30	
	Formic acid	1.5	30	Drained
Washing	Water	400	10	Drained. The leather was hooked to dry ad staked

Note: based on wet blue weight (w/w)

2.4.1 Absorption of the retanning agents.

The content of retanning agents in waste water before and after retanning process were determined by gravimetric method (Shi et al., 2013). The absorption of retanning agents were calculated according to equation (1).

$$\text{Absorption of retanning agents}=\frac{A_1-A_2}{A_1} \qquad (1)$$

Where A_1 is the content of retanning agents in initial liquid and A_2 is the content of retanning agents in liquid after retanning process was completed.

2.4.2 Determination of the absorbance of fatliquoring agent.

In order to explore the effects of the MGE on the absorption of anionic fatliquoring agents, the absorbance of fatliquoring agents were calculated respectively in this experiment according to referenced methods. First, the fatliquoring liquids in initial and finished fatliquoring aqueous solutions were gathered alone, and the fatliquoring contents were tested by TOC. Then, the absorbance of fatliquoring agents were calculated by formula (2).

$$\text{Absorption of fatliquoring}=\frac{B_1-B_2}{B_1} \qquad (2)$$

Where B_1 and B_2 are the contents of fatliquoring in the beginning and finished fatliquoring liquids, respectively.

2.4.3 Determination of the physical and mechanical properties of leathers.

The physical and mechanical properties of leather samples, such as tensile strength, tearing strength were tested by an electronic universal testing machine as per the standard IULTCS(JMV Williams, 2000; J Williams, 2000).

2.4.4 Softness and fullness of leather

The softness of leather samples were measured using a standard method, and their fullness were evaluated by measuring the compressed and resilient thicknesses of leathers according to the method described in literatures(36, 2000; Ding, Yi, Wang, Zhou, & Shi, 2018). In general, higher compressed and resilient thicknesses mean better fullness.

3.Results and discussion

3.1 FT-IR and ^{13}C-NMR of synthesized resin

The chemical structures of melamine (ME), MG and MGE were all characterized by FTIR, as shown in Fig. 2a. The spectrum of ME was observed four characteristic absorption peaks at 3468, 3418, 3332 and 3131 cm^{-1}, which belonged to the stretching vibration absorption of $-NH_2$ groups(Zhu et al., 2020). After the reaction of melamine and glutaraldehyde, the peaks of $-NH_2$ were replaced by a broad absorption peak at 3444 cm^{-1} in the spectrum of MG. This peak was assigned to the stretching vibration of hydroxy groups, proves the successful reaction of melamine and glutaraldehyde(C. Wang, Wu, Li, Mu, & Lin, 2020). The presence of residue aldehyde groups can be confirmed by the peak at 1566cm^{-1}, these aldehydes provide the foundation for subsequent reaction with lysine(Salmani, Abedi, Mozaffari, & Sadeghian, 2017). Additionally, MG exhibited a new absorption peak at 997 cm^{-1} compared with melamine, this peak may attribute to the hydrate of glutaraldehyde. In the FTIR spectrum of MGY, the peak at 997 cm^{-1} disappeared and a new peak at 1625 cm^{-1} assigned to the N–H bending vibration was observed after the addition of lysine, proves the successful reaction between hydroxy and amino groups. After the addition of EDGE, the characteristic peak of epoxide groups at 907 cm^{-1} was almost disappeared and the vibration of ether bond C–O–C at 1179 cm^{-1} and 1051 cm^{-1} were observed, confirming the successful reaction of epoxide groups(Gargol, Klepka, Klapiszewski, & Podkościelna, 2021). Moreover, a characteristic absorption peak of ester groups was observed at 1786 cm^{-1}, implies the reaction between epoxide groups and carboxyl. According to above analysis it is reasonable to believe that melamine, glutaraldehyde, lysine and EDGE have successfully reacted.

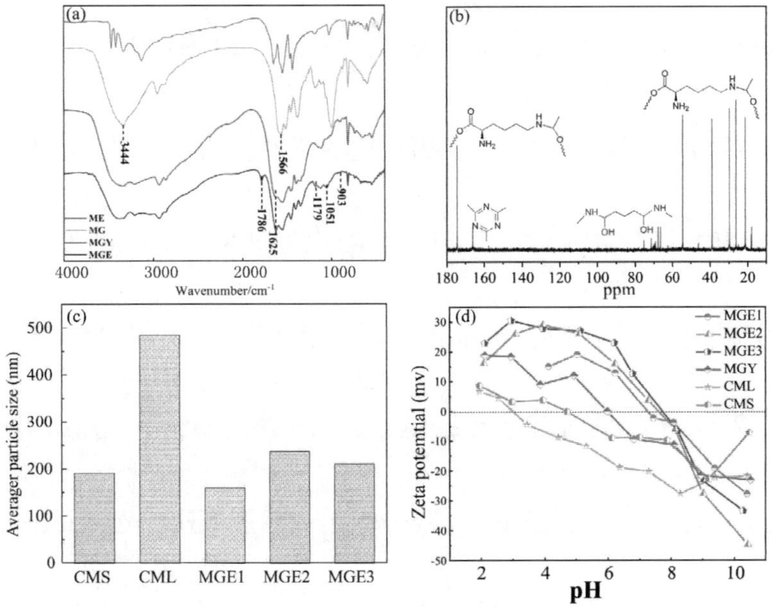

Fig. 2. Characterization of MGE (a) The FT-IR spectra. (b) ^{13}C-NMR of MGE. (c) The average particle size. (d) Zeta potential

The chemical structure of MGE was further characterized by ^{13}C-NMR spectroscopy, as offered in Fig. 2b. First of all, the peak of unreacted melamine at 173 ppm was not observed and the signal observed at 166.1 ppm was assigned to the substituted melamine triazine carbons, indicating the fully reaction of melamine(Deng et al., 2018; Mao, Hassan, & Kim, 2013). Similarly, the absence of the peak at 206 ppm for the carbon of unreacted aldehyde groups in MG also reflects the successful reaction between MG and lysine(No & Kim, 2004). Additionally, the successful introduction of lysine can also be confirmed by the peaks at 17.98-54.37 ppm region, these peaks are attributed to different lysine residues(Eggen & Glomb, 2021; Zhang et al., 2020). Moreover, the 60-80 ppm region were assigned to carbons that linked with the reactions of hydroxyls, these reactions contain the condensation of hydroxyls and the ring-opening reaction between hydroxyls and epoxy groups(Z. Li et al., 2021). Specially, the peak at 62.5 ppm and 69.87 ppm were attributed to the carbon that linked to the ring-opening of epoxy groups and the aliphatic ether bonds of EDGE, respectively. According to literatures, the peak of ether bonds originated from the condensation of hydroxyls should be located at 80-90 ppm, it is notable that this peak was missing in the spectra of MGE, indicating the suppression of methylene-ethers(No & Kim, 2004). Furthermore, the reaction of carboxyl and epoxy group is reinforced by the presence of peak at 174.5 ppm, this peak was assigned to esters generated from the reaction of carboxyl and epoxy groups(Abd-El-Aziz, El-Ghezlani, & Abdelghani, 2020). The higher intensity of this peak suggests that the crosslinking mainly took place between carboxyl and epoxy groups.

3.2 Average particle sizes and zeta potential

The average particle sizes of different resins were determined by Zetatrac particle size analyzer and shown in Fig. 2c. It can be seen that CML exhibited the largest average particle size of 483 nm and the other four resins showed similar average particle sizes of 150-224 nm. Commonly, small particles can easily permeate and further combine with collagen fibers in leather making process. Moreover, benefit from the introduction of lysine, MGE carries abundant functional groups on its main chains. It is advantageous for MGE to form a multi-points interaction with collagen fibers. From the results of particle size testing, it is reasonable to believe that MGE may show good permeability properties in retanning process.

Additionally, electrostatic interaction was viewed as one of the most important reaction types in leather making process. The charge properties of melamine resins at different pH were further analyzed. As can be obviously seen from Fig. 2d, the zeta potential of MGE increased progressively with the increasing dosage of EDGE. Combined with the 13C-NMR analysis results, it can be concluded that the consumption of carboxyl lead to the decreasing of negative charge and finally result in the increase of zeta potential. Moreover, as shown in Fig. 2d, the isoelectric point of MGE3 is 7.55, obviously higher than that of the other two commercial amino resins which are 2.94 and 4.69. Lysine is an alkaline amino acid, and there are large amount of nitrogen atoms served as positive charge centers in the reaction system(Ahmad, Ahmad Kamarudin, Leow, & Rahman, 2020). Moreover, different from traditional synthesis of MR, the sulfonation reaction process was replaced by the introduction of lysine in the preparation process of MGE, resulting the greatly reduction of negative charge centers(X. Wang et al., 2021). In addition, the carboxyl groups will be consumed by the reaction with epoxy groups, the pI of MGE will be inevitably increased. The higher pI is important in leather retanning process, it can improve the binding rates of subsequent anionic fatliquoring agents.

3.3 Retanning behaviors of MGE

3.3.1 Absorptivity of MGE and fatliquoring agents

Fig. 3a shows the absorption rates of leather samples retanned by different melamine resins. It can be obvious seen that the absorption of MGE was up to 77.08%, significantly higher than that of the other resins. It could be suggested that the introduction of lysine in MGE established a biomimetic structure to collagen, which prompt it easily diffuse into collagen fiber system. Moreover, the smaller particle size of MGE also created a favorable condition for its penetration in leather. Additionally, MGE contained abundant functional groups such as hydroxyl, amino and carboxyl, these groups are involved in the formation of interaction among MGE, Cr^{3+} and leather fibers. Ultimately, the strong bond capacity and penetration ability lead to a higher absorption of MGE.

Fig. 3b illustrated the absorption rates of fatliquoring agents. It can be seen that leather retanned by MGE also showed a higher absorption rate of fatliquoring agents than others. As we know, the introduction of amphoteric polymers into retanning process will facilitate the absorption and binding of subsequent materials through the regulation of charge(Hao et al., 2020). As mentioned above, MGE was amphoteric and showed a higher pI, these features can improve its binding ability with collagen fiber and anionic materials, which ultimately leading a higher absorption rate of anionic fatliquoring agents.

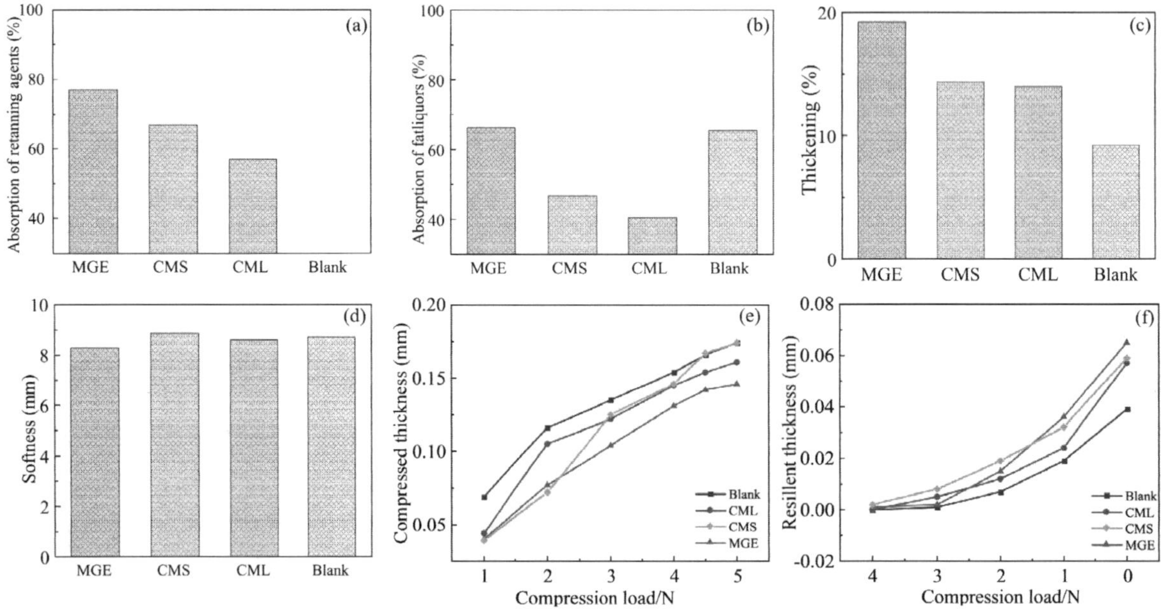

Fig. 3 (a) Absorption of retanning agents. (b) Absorption of fatliquoring agents. (c)Thickening rate. (e)the softness of different leather samples. (f)the fullness of different leather samples

3.3.2 Thickening rates

Thickening rate is an important index to evaluate the performance of melamine resins. Fig. 3c shows the thickening rates of leather retanned by different melamine resins. As illustrated, leather retanned by MGE showed the highest thickening rates of 19.23%, indicating its excellent retanning performance.

3.3.3 The softness and fullness of leather

Fig. 3d shows the softness of different leather samples. It can be seen that the softness of MGE retanned leather decreased to 8.28 mm compared to the blank

(8.7 mm). This result suggested that MGE retanned leather decreased the lubricating degree of collagen fibers, confirming that MGE can endowed leather with higher internal crosslinking intensity(Ding et al., 2020).

Fig. 3e and Fig. 3f shows the fullness of different leather samples. As can be seen, MGE retanned leather showed a better fullness than CMS and CML. This may be attributed to two effects: (1) MGE contains abundant functional groups, these groups make it more effectively to be interlocked in the structures of collagen fibers, thus showed a satisfactory fullness behavior. (2) MGE shows an amphoteric property, it can facilitate the absorption and binding of anionic fatliquors through the regulation of charge, thus leading to a better performance of leather(Hao et al., 2020).

3.3.4 Chromium discharge in effluent

As we know, the chromium that combined with collagen can be released from crust leather in the subsequent processes such as retanning, dyeing, fatliquoring, and so on(Lyu, Chang, Gao, & Ma, 2018). The concentration of chromium in effluents were measured by ICP-AES to evaluate the effects of different melamine resins on the release of chromium, as shown in Fig. 4a. It can be observed that MGE retanned leather showed an obvious lower release of chromium than others. This may attribute to the abundant functional groups of MGE, which can chelate with chromium.

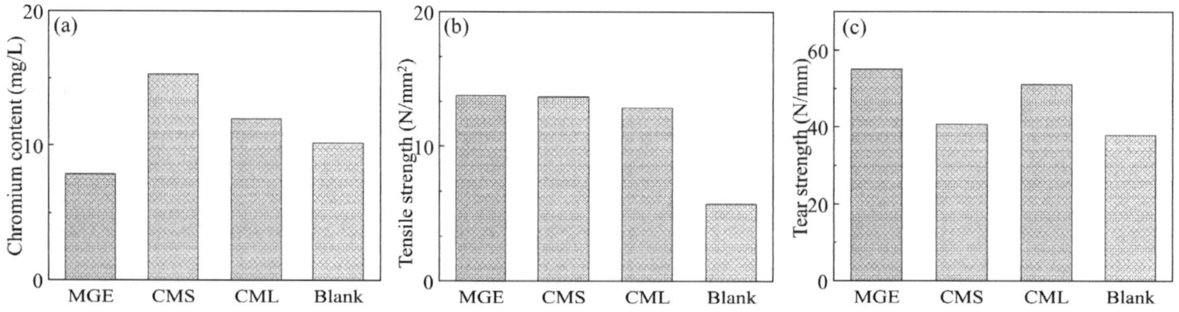

Fig. 4 (a) Chromium discharge in effluents. (b) Tensile strength of leather samples. (c)Tear strength of leather samples

3.3.5 Physical and mechanical properties

Tensile and tear strength tests were carried out for all samples and the results were presented in Fig. 4b and 4c. As can be seen, MGE, CMS and CML showed a similar tensile strength and MGE exhibited obvious higher tear strength. Combined with absorption and thickening results, this result also indicates that MGE can form a multi-points cross-linking structure with collagen fibers. Generally speaking, hydroxyl, carboxyl and amino groups in MGE are all helpful in the interaction with chromium tanned collagen fibers(Lyu, Wang, Ma, Gao, & Jin, 2016). These groups can form hydrogen and coordination bonds with collagen and chromium ions, thus improving the mechanical strength.

4. Conclusion

In this study, melamine was firstly bonded with lysine to improve its amphoteric and water solubility, and a novel formaldehyde-free melamine resin MGE was prepared by epoxy cross-linking strategy. Due to the introduction of lysine, MGE showed an amphoteric property and a high storage stability. Because of the amphoteric property and abundant functional groups, MGE can facilitate the absorption of fatliquoring agents and reduce the release of chromium. These results revealed that the epoxy cross-linking strategy could be a potential alternative for the synthesis of formaldehyde-free melamine resin.

Acknowledgements

This study was financial supported by the National Key R & D Program of China (2017YFB0308500)

Reference

IUP36 (2000) Measurement of leather softness. *Journal of the Society of Leather Technologists and Chemists, 84*, 377-379.

Abd-El-Aziz, A. S., El-Ghezlani, E. G., & Abdelghani, A. A. (2020). Design of Organoiron Dendrimers Containing Paracetamol for Enhanced Antibacterial Efficacy. *Molecules (Basel, Switzerland), 25*(19), 4514. doi:10.3390/molecules25194514

Ahmad, N. N., Ahmad Kamarudin, N. H., Leow, A. T. C., & Rahman, R. N. Z. R. A.

(2020). The Role of Surface Exposed Lysine in Conformational Stability and Functional Properties of Lipase from Staphylococcus Family. *Molecules (Basel, Switzerland), 25*(17), 3858. doi:10.3390/molecules25173858

Akinyemi, B. A., Olamide, O., & Oluwasogo, D. (2019). Formaldehyde free particleboards from wood chip wastes using glutaraldehyde modified cassava starch as binder. *Case Studies in Construction Materials, 11*, e00236. doi:https://doi.org/10.1016/j.cscm.2019.e00236

Arias, A., González-García, S., Feijoo, G., & Moreira, M. T. (2021). Environmental benefits of soy-based bio-adhesives as an alternative to formaldehyde-based options. *Environmental Science and Pollution Research, 28*(23), 29781-29794. doi:10.1007/s11356-021-12766-4

Balducci, F., Adamopoulos, S., Pettinari, C., Canti, E., Di Nicola, C., Tombesi, A., . . . Gabbani, C. (2020). A Formaldehyde-free Adhesive for Particleboards Based on Soy Flour, Magnesium Oxide, and a Plant-derived Enzymatic Hydrolysate. *2020, 15*(2), 16. Retrieved from https://ojs.cnr.ncsu.edu/index.php/BioRes/article/view/BioRes_15_2_3087_Balducci_Formaldehyde_Free_Adhesive_Particleboards

Cai, H., Wang, Z., Xie, D., Zhao, P., Sun, J., Qin, D., & Cheng, F. (2021). Flexible transparent wood enabled by epoxy resin and ethylene glycol diglycidyl ether. *Journal of Forestry Research, 32*(4), 1779-1787. doi:10.1007/s11676-020-01201-y

Deng, S., Pizzi, A., Du, G., Lagel, M. C., Delmotte, L., & Abdalla, S. (2018). Synthesis, structure characterization and application of melamine–glyoxal adhesive resins. *European Journal of Wood and Wood Products, 76*(1), 283-296. doi:10.1007/s00107-017-1184-9

Ding, W., Pang, X., Ding, Z., Tsang, D. C. W., Jiang, Z., & Shi, B. (2020). Constructing a robust chrome-free leather tanned by biomass-derived polyaldehyde via crosslinking with chitosan derivatives. *Journal of Hazardous Materials, 396*, 122771. doi:https://doi.org/10.1016/j.jhazmat.2020.122771

Ding, W., Yi, Y., Wang, Y.-n., Zhou, J., & Shi, B. (2018). Preparation of a highly effective organic tanning agent with wide molecular weight distribution from bio-renewable sodium alginate. *ChemistrySelect, 3*(43), 12330-12335.

Eggen, M. D., & Glomb, M. A. (2021). Novel Amidine Protein Cross-Links Formed by the Reaction of Glyoxal with Lysine. *Journal of Agricultural and Food Chemistry, 69*(28), 7960-7968. doi:10.1021/acs.jafc.1c02792

Gargol, M., Klepka, T., Klapiszewski, Ł., & Podkościelna, B. (2021). Synthesis and

Thermo-Mechanical Study of Epoxy Resin-Based Composites with Waste Fibers of Hemp as an Eco-Friendly Filler. *Polymers, 13*(4). doi:10.3390/polym13040503

Hao, D., Wang, X., Liu, X., Zhu, X., Sun, S., Li, J., & Yue, O. (2020). A novel eco-friendly imidazole ionic liquids based amphoteric polymers for high performance fatliquoring in chromium-free tanned leather production. *Journal of Hazardous Materials, 399*, 123048. doi:https://doi.org/10.1016/j.jhazmat.2020.123048

Ji-bo, Z., Jian-fei, Z., Pei-lin, L., Xue-pin, L., & Bi, S. (2018). Preparation of formaldehyde-free melamine resin using furfural as condensation agent and its retanning performances investigation. *Journal of the American Leather Chemists Association, 113*(6).

Kan, Y., Sun, B., Kan, H., Bai, Y., & Gao, Z. (2021). Preparation and characterization of a melamine-urea-glyoxal resin and its modified soybean adhesive. *International Journal of Adhesion and Adhesives, 111*, 102986. doi:https://doi.org/10.1016/j.ijadhadh.2021.102986

Kord, B., Movahedi, F., Adlnasab, L., & Ayrilmis, N. (2021). Decreasing Formaldehyde Emission from Particleboard Panels Fabricated by Adding Newly Phenolic Compounds as a Scavenger to Urea Formaldehyde Resin.

Li, J., & Zhang, Y. (2021). Morphology and Crystallinity of Urea-Formaldehyde Resin Adhesives with Different Molar Ratios. *Polymers, 13*(5), 673. Retrieved from https://www.mdpi.com/2073-4360/13/5/673

Li, Z., Zhi, Y., Ni, Y., Su, H., Miao, Y., & Shan, S. (2021). Novel melamine-based porous organic materials as metal-free catalysts for copolymerization of SO2 with epoxide. *Polymer, 217*, 123434. doi:https://doi.org/10.1016/j.polymer.2021.123434

Liu, X., Wang, W., Wang, X., Sun, S., & Wei, C. (2021). A "Taiji-Bagua" inspired multi-functional amphoteric polymer for ecological chromium-free organic tanned leather production: Integration of retanning and fatliquoring. *Journal of Cleaner Production, 319*, 128658.

Lyu, B., Chang, R., Gao, D., & Ma, J. (2018). Chromium Footprint Reduction: Nanocomposites as Efficient Pretanning Agents for Cowhide Shoe Upper Leather. *ACS Sustainable Chemistry & Engineering, 6*(4), 5413-5423. doi:10.1021/acssuschemeng.8b00233

Lyu, B., Wang, H.-D., Ma, J.-z., Gao, D.-g., & Jin, P. (2016). Preparation and application of castor oil/nano-TiO2 composite fatliquoring agent via a

Pickering emulsion method. *Journal of Cleaner Production, 126*, 711-716. doi:https://doi.org/10.1016/j.jclepro.2016.02.099

Mao, A., Hassan, E. B., & Kim, M. G. (2013). The effects of adding melamine at different resin synthesis points of low mole ratio urea-melamine-formaldehyde (UMF) resins. *BIORESOURCES, 8*(4), 5733-5748. doi:10.15376/biores.8.4.5733-5748

No, B. Y., & Kim, M. G. (2004). Syntheses and properties of low-level melamine-modified urea–melamine–formaldehyde resins. *Journal of Applied Polymer Science, 93*(6), 2559-2569. doi:https://doi.org/10.1002/app.20778

Okumura, H. (2021). Application of phenolic compounds in plants for green chemical materials. *Current Opinion in Green and Sustainable Chemistry, 27*, 100418. doi:https://doi.org/10.1016/j.cogsc.2020.100418

Salmani, M. H., Abedi, M., Mozaffari, S. A., & Sadeghian, H. A. (2017). Modification of pomegranate waste with iron ions a green composite for removal of Pb from aqueous solution: equilibrium, thermodynamic and kinetic studies. *AMB Express, 7*(1), 225-225. doi:10.1186/s13568-017-0520-0

Song, J., Chen, S., Yi, X., Zhao, X., Zhang, J., Liu, X., & Liu, B. (2021). Preparation and Properties of the Urea-Formaldehyde Res-In/Reactive Halloysite Nanocomposites Adhesive with Low-Formaldehyde Emission and Good Water Resistance. *Polymers, 13*(14), 2224. Retrieved from https://www.mdpi.com/2073-4360/13/14/2224

Wang, C., Wu, J., Li, L., Mu, C., & Lin, W. (2020). A facile preparation of a novel non-leaching antimicrobial waterborne polyurethane leather coating functionalized by quaternary phosphonium salt. *Journal of Leather Science and Engineering, 2*(1), 2. doi:10.1186/s42825-019-0014-8

Wang, J. (2021). Synthesis and Computer Simulation Analysis of MF Modified Starch Adhesive. *Journal of Physics: Conference Series, 1992*(2), 022171. doi:10.1088/1742-6596/1992/2/022171

Wang, X., Sun, S., Zhu, X., Guo, P., Liu, X., Liu, C., & Lei, M. (2021). Application of amphoteric polymers in the process of leather post-tanning. *Journal of Leather Science and Engineering, 3*(1), 9. doi:10.1186/s42825-021-00050-7

Williams, J. (2000). IULTCS (IUP) test methods-Measurement of tear load-double edge tear. *Journal of the Society of Leather Technologists and Chemists, 84*, 327-329.

Williams, J. (2000). IULTCS (IUP) test methods-Measurement of tensile strength and percentage elongation. *Journal of the Society of Leather Technologists and*

Chemists, 84, 317-321.

Zhang, L., Yan, P., Li, Y., He, X., Dai, Y., & Tan, Z. (2020). Preparation and antibacterial activity of a cellulose-based Schiff base derived from dialdehyde cellulose and L-lysine. *Industrial Crops and Products, 145*, 112126. doi:https://doi.org/10.1016/j.indcrop.2020.112126

Zhu, Z., Lin, P., Wang, H., Wang, L., Yu, B., & Yang, F. (2020). A facile one-step synthesis of highly efficient melamine salt reactive flame retardant for epoxy resin. *Journal of Materials Science, 55*(27), 12836-12847. doi:10.1007/s10853-020-04935-6

PREPARATION OF SHOE SOLING MATERIAL FROM LEATHER AND TEXTILE WASTE

Sanjeev Gupta[1], S. Ponsubbiah[1a], R.Murugesan[2]& Aman Malhotra[3]

1. *CSIR– Central Leather Research Institute, Chennai,India.*

1a. Institute of Leather Technology, Chennai, India

2.Anna University,Chennai,India

3. Umeå University, Sweden

1.Corresponding author: sgupta1281@gmail.com

Abstract

The aim of this research is recycling of the leather and textile industry solid waste such as chrome shaving and weaving mills hard waste into composite material, such as shoe soling. Chrome shaving and weaving waste from the leather and textile industry were used for making composites by mixing with rubber. Isoprene and Ethylene propylene monomer (EPDM) rubbers were used for this purpose. Various combinations of rubber, chrome shaving waste and weaving hard waste were studied to get the desired characteristics of shoe soling material. The prepared composite materials were characterized for physico-mechanical behaviours like hardness, density, abrasion resistance and tensile strength, and compared with those of the rubber-based soling material available commercially. The composite prepared using a combination of the isoprene and EPDM (70 Phr :30 Phr) rubber along with chrome shaving and weaving hard waste showed physico-mechanical characteristics very close to the commercial available material with higher abrasion resistance. Hence, an efficient use of the fibrous chrome shaving and weaving wastes from leather and textile industry in such composite making would avoid the environmental problem, and could be a source of sustainable value-creation.

Figure 1 –Flow diagram of this research work

1 Introduction

One of the hardest questions of today comes to mind at this point: How will we manage the resultant waste of the products we created? Should we dump the waste to proper waste yards and reuse or recycle them? The answers to these questions are crucial. Scientists, governments, and local authorities work for finding answers to these questions. Wastes can be a problem for local authorities because of their environmental effects.

The waste generated from the leather industry is unavoidable. Liquid and solid wastes are generated during the processing of Hides and Skins. There are noticeable progress are present to treat liquid waste. An easy way to dispose solid wastes like chrome shavings and trimmings is land filling, which is subjected to strict environmental protocols due to the presence of chromium in these solid wastes. In India, more than 24million pieces of cattle hides, 22million of buffalo hides, 106million of goat skins and 37million of sheep skins are processed in about 1600 tanneries of leather industry. This generates around 2 million tons of solid wastes annually (CLRI, 2015). Chrome shavings are considered as a hazardous waste owing to the chrome present in it, also there is a certain risk that chromium may be washed off and it may contaminate the ground water if dumped carelessly. Hence, the safe disposal/application of chrome shavings should be addressed. Presently, chrome shavings are utilized as several low value applications such as –

- Preparation of regenerated leather using natural and synthetic blend polymer (US Patent)(Parrini et.al. 1979)
- Composites made from chrome shavings and natural fibres using rubber solution (US Patent)(Parrini et. al. 1979)
- A composite of natural and synthetic fibres as a leather substitute (US Patent) (Parrini et. al. 1979)
- Production of parchment like materials (Indian Patent No.300DEL 1999)
- Value added composite from leather and no-leather fibres (US Patent) (Parrini et. al. 1979)
- Sustainable Landfill (Indian Patent No.125DEL 2006).

However, most of these applications have not been commercialized. Conventional leather boards have also been made that are used as in Composites and packing materials etc.The products in cotton textiles, from registered manufacturing sector in India, have yielded an output value of Rs. 9,714 crore in 1981-82. This has risen to Rs. 29,746 crores in 1990-91, which reflects the degree of expansion. The cotton textile industry including ginning units, spinning units and weaving & processing, produce wastes, which have intrinsic value of almost Rs. 100 million per year. This relates to only one type of solid waste of textile industry. It is estimated that the total solid wastes, including those of man-made fibres, and rags which are collected in huge quantities, when recycled would amount to Rs. 300 million per year. The other wastes in the textile industries, for example chemicals and dyes have been estimated between Rs. 300 to 350 million per year. There are about 270 registered composite mills in our country.

Klein classified cotton fibers used in short-staple spinning as virgin fiber (from ginning mill), clean waste, comber waste, recycled fibers from dirty waste, and fibers torn out of hard waste (roving, yarn, and twisted threads) .. Broken ends of sliver, lap, web, and filter strippings from draw frame, roving frame, ring spinning frame, and rotor spinning frame are known as clean waste, having more than 95% of good fiber. Comber and roving wastes' good fiber ratio is around 95–97%. Wastes generated in blowroom machines, and cards are dirty wastes with 35–55% good fiber ratio. Besides, another dirty waste, flat and filter stripping waste, has a higher amount of good fiber (65–80%). As the waste fibers are processed in different number of machines and therefore stressed fibers, their good fiber content is less than virgin fibers. For this reason, spinners prefer to feed the waste fibers into normal spinning process, in a controlled manner, with a constant percentage in order to avoid quality variations. Generally, wastes arising in the mills can be returned to the same blend from which they arose; comber wastes are mostly used in rotor spinning. In carded ring-spun yarn and fine rotor-spun yarn production, waste fibers can be used, up to 5%, but for combed yarns, waste fiber ratio can be lower, up to 2.5%. Higher waste fiber amounts can be used for medium and coarser rotor yarns, about 10 and 20%, respectively.

Several types of wastage are found during the processing of row cotton in spinning mill 12-13% wastage are produced for card process and 25-30% wastage are produced for combed process. The names of different wastage are given below

S.No	Waste Generated from various unit Operations	Sub Classifications
1	Blow-room of waste = [3%]	i. Droppings-2 ii. Lap waste.
2	Carding of waste = [3-5%]	i. Droppings-1 ii. Sliver waste iii. Vaccum waste.
3	Draw frame of waste = [0.5%]	i. Sliver waste.
4	Lap former of waste = [0.5%]	i. Lap waste ii. Sliver waste iii. Vaccum waste.
5	Comber of waste = [12-18%]	i. Lap waste ii. Noils iii. Vaccum waste.
6	Simplex of waste = [0.5%]	i. Roving waste ii. Sliver waste
7	Ring frame of waste = [2-3%]	i. Pneumafil ii. Hard waste iii. Roller lapping iv. Sweeping v. Vaccum waste
8	Winding section of waste = [1%]	i. Hard waste.

Hard waste

The hard waste occurs mostly at the final stage, ie, in post spinning process (Cone winding machine). These hard wastes are not reusable in spinning industry.

- The hard waste level obtained in cone winding machine ranges from 0.3 to 0.8 per cent.
- The hard waste level obtained in warping machine ranges from 0.1 to 0.4 percent
- The hard waste level obtained in weaving machine ranges from 0.1 to 0.6 percent

Since huge amount of yarns are processed through winding, warping and weaving machine, the quantity of hard waste generated every day also higher. Nearly 100kg of hard wastes are generated every day in a textile composite mill.

Major reasons for the high hard waste at cone winding machine are

- Inadequate maintenance of machinery
- Clearer cuts in automatic cone winders
- Cop rejection
- Repeaters
- Wrong work practices in Ring frame and cone winding
- Materials handling & housekeeping

The technologies developed by various research institutions in our country have been implemented in various textile mills which have resulted in pollution control and biogas generation.

Table 1 – Characterisation of waste generated from spinning mills

Waste type	Fibre content (%)	2,5% SL* (HVI) (upper length) (mm)	50% SL (mm)	ML* (AFIS) (LCT length) (mm)	Fibre hooks (%)	SFL* (mm)	SFA* (%)	SFC* (AFIS short fibre amount) (%)	Staple gradient (HVI uniformity ratio)
Blow room	45.74	25.27	9.24	15.62	5.3	7.26	56.4	38.0	38.4
Card	82.44	21.69	6.71	7.79	1.2	2.81	78.0	76.6	33.0
Sliver	82.11	27.63	11.62	11.04	0.4	11.04	44.5	15.0	42.0
Fabric	80.89	22.64	8.46	13.36	8.4	6.07	57.2	47.6	39.6
Weaving	99.41	27.67	12.50	20.59	4.8	10.38	42.5	17.4	42.5

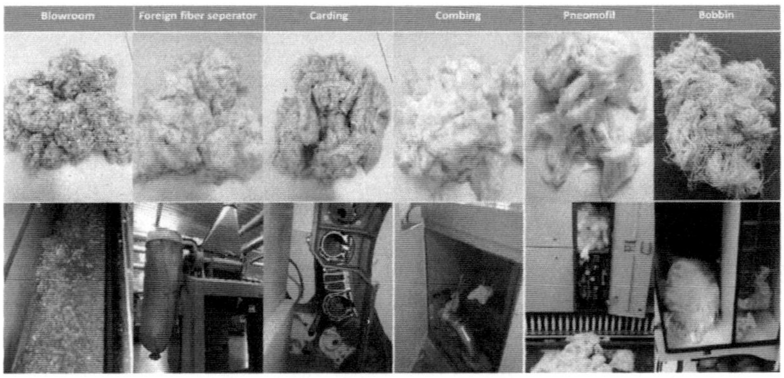

Figure 2 –Various waste generated from the spinning mills

The present communication describes the making of composite materials using chrome shaving waste and weaving mill waste as the filler. Besides additive material was also used along with the chrome shaving and hard weaving mill waste. The influence of chrome shaving waste and hard weaving waste with rubber material on the physico-mechanical properties of the composite were studied and compared with those properties of commercial products.

2 Materials and Methods

2.1 Materials

Isoprene rubber, EPDM rubber were used for preparation of composite materials, Zinc oxide and Stearic Acid as an activator &Pilcure CBS, Pilcure MBTS &Pilcure TMT as an accelerator and other chemicals like sulphur used is laboratory grade. Chrome shaving waste was collected from the leather industry and the spinning wastes are collected from spinning mill.

2.2 Pre-treatment of the Chrome shavings

Normally chrome shavings are acidic in nature. To remove the acidity, the chrome shaving wastes collected from the leather industry were treated with aqueous ammonia, 1% solution of urea and sodium bicarbonate followed by washing with water and drying under sunlight for 24 hours. The dried chrome shaving fibres were turned into fine particles by using strap cutting machine and stored for further use.

2.3 Pre-treatment of the hard weaving millwaste

The collected hard weaving mill wastes are washed with water and soap oil for 2 hours and drying under sunlight for 24 hours. The dried hard weaving mill waste were turned into fine particles by using strap cutting machine and stored for further use.

2.4 Preparation of soling material

Rubber Composites of six different compositions were made using isoprene and/or EPDM rubber along with the treated chrome shaving and hard weaving mill waste as filler. The detail compositions are presented in Table 1 and Table 2. Table 1 presents the three different compositions of rubber Composites, TA, TB and TC. Table 2 presents the sole compositions similar that of TC but the Isoprene and EPDM rubber proposition is changed as 60/40,70/30 & 80/20.

Table 2: Trial with Isoprene & EPDM Rubber and chrome shavings

Composition of the different Trials			**Role of the material**	**Quantity (Phr)**
Trial-A (TA)	**Trial-B (TB)**	**Trial-C (TC)**		
Isoprene rubber	EPDM Rubber	Isoprene + EPDM rubber	Base polymer	100
Chrome shaving	Chrome shaving	Chrome shaving	Solid waste / filler	25
Weaving hard waste	Weaving **hard** waste	Weaving **hard** waste	Solid waste / filler	25
Zinc oxide	Zinc oxide	Zinc oxide	Activator	10
Stearic acid	Stearic acid	Stearic acid	Activator	4
CBS	CBS	CBS	Accelerator	1

MBTS	MBTS	MBTS	Accelerator	1
TMT	TMT	TMT	Accelerator	0.5
Sulphur	Sulphur	Sulphur	Vulcanising Agent	5
$CaCO_3$	$CaCO_3$	$CaCO_3$	Filler	10

Table 3: Trial with Isoprene & EPDM Rubber and chrome shavings

Composition of the different Trials			**Role of the material**	**Quantity (Phr)**
TD	**TE**	**TF**		
Isoprene rubber	Isoprene rubber	Isoprene rubber	Base polymer	60/70/80
EPDM Rubber	EPDM Rubber	EPDM Rubber	Base polymer	40/30/20
Chrome shaving	Chrome shaving	Chrome shaving	Solid waste / filler	25
Weaving hard waste	Weaving hard waste	Weaving hard waste	Solid waste / filler	25
Zinc oxide	Zinc oxide	Zinc oxide	Activator	10
Stearic acid	Stearic acid	Stearic acid	Activator	4
CBS	CBS	CBS	Accelerator	1
MBTS	MBTS	MBTS	Accelerator	1
TMT	TMT	TMT	Accelerator	0.5
Sulphur	Sulphur	Sulphur	Vulcanising Agent	5
CaCo3	**CaCo3**	**CaCo3**	**Filler**	**10**

To prepare the sole samples, accurately weighed quantities of the ingredients were fed to a two roll rubber mixing mill (with roller dimension D=220mm and L = 450 mm) maintained at a temperature of 47-67 °C and rolling at a speed of 15 rpm. Composite of 5–6 mm thickness were made and stored at room temperature.

The above experiments were carried out in the curing temperature around 140–160 ºC and curing time around 5 to 12 minutes in the compression moulding machine. Corresponding female and male mould halves were used in this process. A pre-weighed charge cut were placed inside the mould. To facilitate polymerization (or cross-linking) and consolidation of composite material an appropriate pressure force were maintained. During this compounding process proper nip cap and cuts were maintained in order to get the uniform compounding.

Composites of different compositions, TA, TB, TC, TD, TE and TF were made following the above mentioned procedure.

2.3 Characterization techniques

The morphology of the prepared composite materials was characterized using Scanning Electron Microscopy (SEM make: Phenom world, model: phenom Pro). Physico-mechanical properties of the sole materials were studied using Universal Test Machine (INSTRON, model: 3369/J7257), Bata Flexing Resistance (SATRA, Model: STM 612) and Leather Sole Abrasion Tester (SATRA, Model: STM 140).

3 Results and Discussions

3.1 Scanning Electron Microscopic (SEM) Study

The SEM Micrographs of the fractured cross sections of the tensile tested specimens of TA, TB,TC,TD,TE and TF are presented in figure 1.

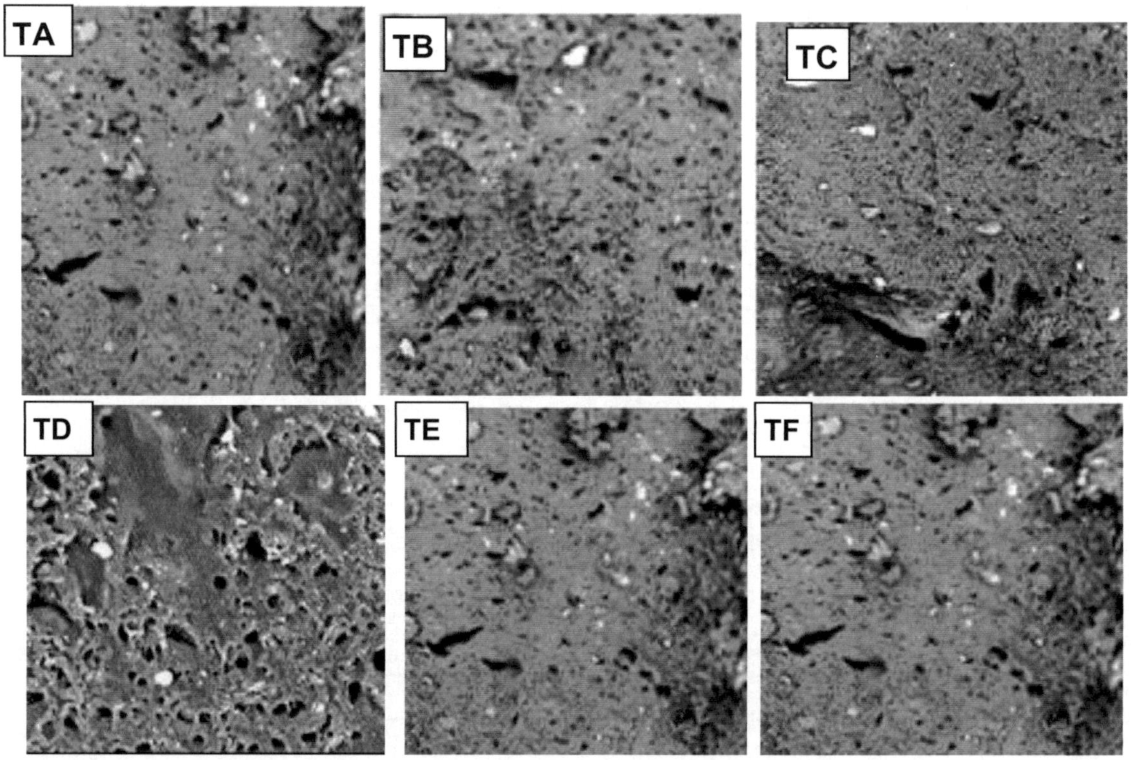

Figure 3: SEM Images of Trial TA, TB, TC, TD, TE & TF

Above SEM images of the fractured cross sections of the tensile tested specimens of TA, TB,TC,TD,TE and TF. The SEM images of the experiment I and experiment II are shown in the above Fig. 3. The SEM images of sample TA, TB and TC show aggregation of fibre due to blending of Isoprene and EPDM with chrome shavings and weaving hard waste. The SEM image reveals that in all three samples chrome shavings and weaving hard wastes are intermingled with polymers and also can be seen that in all the six samples chrome shavings and weaving hard wastes are closely knitted.

3.2 Physico-Mechanical properties

The prepared soling materials were subjected to physic-mechanical testing to study the properties like, abrasion resistance, tensile strength, percentage of elongation at break, hardness, density and flexing endurance. The values obtained for these properties were compared with those properties of the control sole sample. The control sole sample is a commercial rubber sole obtained from AGGU Composites Pvt Ltd, Ranipet, Tamilnadu, India. The physico-mechanical properties of the control sole material are tabulated below:

Table 3: Values of physico-mechanical parameters of the control sole sample(as per SATRA standard).

Sr.No.	Mechanical Properties	Values
1.	Tensile strength	8.0 MPa
2.	Percentage of Elongation	200 %
3.	Flexing endurance	50000 flexes
4.	Density	1.25 gm/cc
5.	Abrasion Resistance	150 mm^3
6.	Hardness	65 (Shore A)
7.	Thickness	0.6 mm

The physico-mechanical characteristics of the rubber Composites made using isoprene and/or EPDM rubber along with chrome shaving waste (as per the composition listed in table 1) are presented in table 4. The same characteristics of the rubber Composites made using mixture of isoprene and EPDM rubber along with chrome shaving waste and hard weaving waste with inorganic nanomaterials as filler (as per the compositions in table 2) are presented in table 5.

Table 4: Values of physico-mechanical parameters of the rubber sole samples prepared using chrome shaving waste as filler.

Sole Sample	Tensile Strength (MPa)	Elongation at Break (%)	Flexing endurance (flexes)	Density (gm/cc)	Abrasion Resistance (mm^3)	Hardness (Shore A)
Control	8.0	200	50455	1.16	148.22	65
TA	6.0	205	48220	1.01	120.98	62
TB	6.5	170	35500	1.37	395.20	67
TC	7.4	180	46225	1.24	290.25	65

The values presented in Table 4 show that the properties like, sole flexing (except TB), density, and hardness, of the sole samples prepared by using chrome shaving waste and weaving hard waste as fillers (TA, TB and TC) are comparable with those of the control sole sample. The abrasion resistance values of these Composites are much higher than that of the control sample. However, the values of tensile strength are lower (almost half) than that of the control sample.

Based on the physico-mechanical properties of the sole samples TA, TB and TC, the sole composition of TC was selected for further study. The another experiment were carried out by different proposition of Isoprene and EPDM rubber(60/40,70/30 &80/20) in order to achieve the desirable properties of soling material.

Table 5: Values of Physico-mechanical parameters of the rubber sole samples prepared using chrome shaving and weaving hard waste along with different proposition of Isoprene and EPDM rubber.

Sole Sample	Tensile Strength (MPa)	Elongation at Break (%)	Flexing endurance (flexes)	Density (gm/cc)	Abrasion Resistance (mm^3)	Hardness (Shore A)

Control	8.0	200	50455	1.16	148.22	65
TD	7.8	175	48523	1.25	310.00	67
TE	7.7	198	49056	1.20	302.25	66
TF	7.5	210	49876	1.18	300.45	65

The values of physic-mechanical characteristics presented in table 5 clearly show that increasing the proposition of isoprene along with the chrome shaving waste and weaving hard waste improves all the sole characteristics. All the sole samples in this trial showed almost similar tensile strength. Among the three trials(TD,TE & TF),the sole prepared by 70-Phr of Isoprene and 30-Phr of EPDM rubber showed excellent soling characteristics when compared with the control sole sample.

When we compare the final results, it is clear that Chrome shavings and weaving hard waste have good compatibility between Isoprene and EPDM rubbers. There is a gradual increase in Tensile strength when the isoprene rubber proposition incresed in this experiment. The maximum tensile strength obtained when the 70 Phr of isoprene and 30 Phr EPDM Rubber added with the chrome shavings and textile hard waste in this composition. This value is almost equal to the requirement of shoe soling material. The rest of the materials are shown slightly lesser tensile properties when compare to this composition. Elongation at break is almost same as using $CaCO_3$ filler in this composition. The value is between 192% to 205%, but the requirement is 150 % (control).Hence 70 Phr of Isoprene,30Phr of EPDM and 25 Phr of Chrome shaving with 25 Phr of weaving hard waste composition is consider for further developments.

4. Conclusion

Composite samples were successfully made using the chrome shaving waste and hard weaving waste with rubber and additives. The SEM images showed that the chrome shaving waste and hard weaving wasteshowed good compatibility with both isoprene and EPDM rubber. The samples with lesser proposition of EPDM rubber showed good physico-mechanical characteristics. However, incorporation of Isoprene and EPDM rubber along with the solid waste improved the tensile strength of the samples. Therefore, the 70-Phr of isoprene + 30-Phr of EPDM + 50-Phr of Leather and textile waste composite can be considered for further development, which in turn could be a sustainable way of making use of the hazardous solid waste generated by the leather and textile industry. Hence, it may be concluded that efficient use of the fibrous nature of waste from textile and Leather industry would avoid the environmental problem, and could be a source of sustainable value-creation from these solid wastes.

References

1. Alexender, K.T.W.; Corning, D.R.; Cory, N.J.; Donohue, V.J. & Sykes, R.L. – J Soc., Leather Trchnol. Chem., 76, 17 (1992).
2. A Bartl.,Textile waste,Waste,Chapter 12,2011-Academic Press, Pages 167-179.

3. Anoop V, Subramani Sankaraiah, S.N. Jaisankar, Sohini Chakraborthy & Mary N.L. Enhanced mechanical, thermal and adhesion properties of polysilsesquioxane speres reinforced epoxy nanocomposite adhesive, J.Adhes., 2019, DOI: 10.1080/00218464.2019.1620107.
4. B.Rajeswari, N.Malarvizhi, E.Deenadayalan and Sellamuthu N. Jaisankar "Influences of functionalized nanoclays on morphology and morphology and mechanical properties of polyvinyl alcohol based composites by twin-screw extruder, Polym. Plast. Technol. Eng. 56, 883-888, 2017.
5. Bauer, D.H.; Donrad, E.T. & Lofy, R.J. – Am. Leather Chem. Technol, Ass., 72, 270 (1977).
6. Burgoyne, M; Leaker, G. & Krekic, Z – Rubber Chem, Technol., 49, 375 (1976)
7. Dhanasekaran Prakash and Sellamuthu N. Jaisankar, Thermoplastic poly (urethane-thiourethane) triblock copolymers with SWCNTs composite, Diam. Relat. Master. 93, 34-41, 2019.
8. Flory, P.H. – J. Chem. Phys., 18, 108 (1963)
9. He Xianqi S.B. & Haslam E., Gelatin – Polyphenol interaction, J Am Leather Chem Assoc, 89 (1994) 98-104.
10. Kremem, S.S. & Lollar, R.M. – J. Am. Leather Chem. Ass., 46, 34 (1951)
11. Marvin, M. – Rubber Chem. Technol., 75, 429 (2002).
12. Natchimuthu, N., Radhakrishnan, G., Palanivel, K., Ramamurthy, K. and Anand, J.S. (1994). Vulcanization Characteristics and Mechanical Properties of Nitrile Rubber Filled with Short Leather Fibres, Polymer Int, 33: 329-333.
13. Parvathi M.S., Suseela. K. & Nandy S.C., Chromium and leather wastes, Leath sci. 33 (1986) 303-309.
14. Przepiorkowska, A., Prochon, M. amd Zaborski, M. (2004). Use of Waste Collagen as a Filler for Rubber Compounds, J Soc. Leather Technol. Chem, 88: 223-227.
15. Ravichandran, K and Natchmuthu, N. (2005). Vulconization Characteristics and Mechanical Properties of Natural Rubber-Scrap Rubber Compositions Filled with Leather Particles, Polymer Int., 54: 553-559.
16. Ravichandran, K. & Natchimuthu, N. – Int. Sem. On Adv. In Pol. Technol., Cochin India, 25 (2002).
17. Ravichandran, K. and Natchimuthu, N. (2005). Natural Rubber-Leather Composites, Polimeros: Ciencia e Tecnologia, 15: 102-108.
18. Report of the meeting of the IULTCS Tannery wastes commission (1981). Waalwijk, Holland 9-11 May. T.P.Sastry, P.K.Sehal and T.Ramasami: Value added eco friendly products from tannery solid wastes. Journal of Environmental science & engineering.
19. Seena Joseph, Tushar S., Ambone, Abhijit V. Salvekar, S.N.Jaisankar, P.Saravanan, E.Deenadayalan, Processing and characterization of waste leather based ploycaprolactonebiocomposites, Polym. Composite (DOI: 10.1002/pc.23891) (in press 2015).
20. Sellamuthu N.Jaisankar, Dorina J., Nelson and Christoper Brammer, Preparation and properties of thermoplastic nanocomposites based on polyurethane ionomers, Proc. of 53rd Pentasectional Meeting of the American Chemical Society, 14, 2008.
21. Srinivasan T.S., M.Anendrakumal: Krishnan T.S. and K.J.Scaria (1985) "Chrome Shavings – a tannery waste, current practice and future trends for its utilization" Presented in the 3rd AAP Animal Science Congress, Seoul, Korea.
22. Srinivasan, T.S. (1980). A sample survey conducted on availability of tannery solid waste in Tamilnadu. Balachandaran unni Nair, Raghava Rao. J, Sreeram K.J. and Thanikaivelan. P., (2002)

"Green Route for the utilization of chrome shavings" Environmental science and Technology vol. 36 No.6pp 1372-1376.

23. Suseela K, Parvathi M.S. & Nandy S.C., Chromium containing leather wastes, Leder, 37 (1986) 45-47.
24. Y Wang .,Fiber and Textile waste utilisation, Waste and biomass valorization,2010-Springer-1,135-143

IULTCS Congress
3rd – 5th November 2021

PATHWAY TO A SUSTAINABLE LEATHER INDUSTRY

Deborah Taylor
Managing Director
Sustainable Leather Foundation

SUSTAINABLE LEATHER FOUNDATION

If you take away just one thing from this presentation, I hope it is this:

If we want our leather industry to be sustainable for future generations we have to act now!

SUSTAINABLE LEATHER FOUNDATION

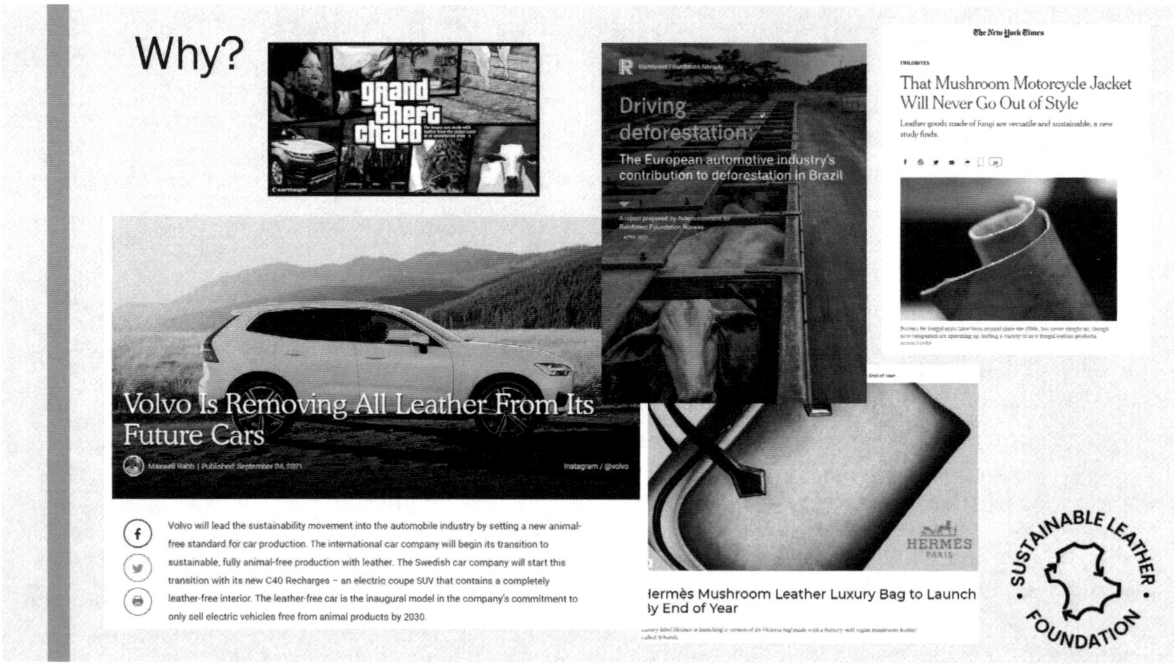
Why?
grand theft chaco
Driving deforestation:
The European automotive industry's contribution to deforestation in Brazil
The New York Times
That Mushroom Motorcycle Jacket Will Never Go Out of Style
Leather goods made of fungi are versatile and sustainable, a new study finds.
Volvo Is Removing All Leather From Its Future Cars
Instagram / @volvo
Volvo will lead the sustainability movement into the automobile industry by setting a new animal-free standard for car production. The international car company will begin its transition to sustainable, fully animal-free production with leather. The Swedish car company will start this transition with its new C40 Recharges – an electric coupe SUV that contains a completely leather-free interior. The leather-free car is the inaugural model in the company's commitment to only sell electric vehicles free from animal products by 2030.
HERMES PARIS
lermès Mushroom Leather Luxury Bag to Launch
3y End of Year
SUSTAINABLE LEATHER FOUNDATION

It's not all bad news....
First Zimbabwe tanners join Sustainable Leather Foundation
Worldwide
Published: 17 August, 2021
Two tanneries from Zimbabwe have become the first from the country to join the Sustainable Leather Foundation (SLF).
Credit: Sustainable Leather Foundation
JAGUAR LAND ROVER TRIALS WORLD-FIRST DIGITAL SUPPLY CHAIN FOR LEATHER USING BLOCKCHAIN TECHNOLOGY
13 OCTOBER 2021
PFC FREE: GETTING 'FOREVER CHEMICALS' OUT OF FOOTWEAR
0% TOXIC PFCS
FINANCIAL TIMES
A new world is possible.
Letter: Leather industry could lead the way in British manufacturing
SUSTAINABLE LEATHER FOUNDATION

How Can We Sustainably Future Proof Our Industry?

- Invest in:
  - *People*
  - *Research & Development*
  - *Collaboration*
  - *Industry Support Associations*
- Stop:
  - *Scoring points off fellow industry actors*
  - *Hiding behind "tradition"*
  - *Comparing apples and oranges*

SUSTAINABLE LEATHER FOUNDATION

Create your pathway

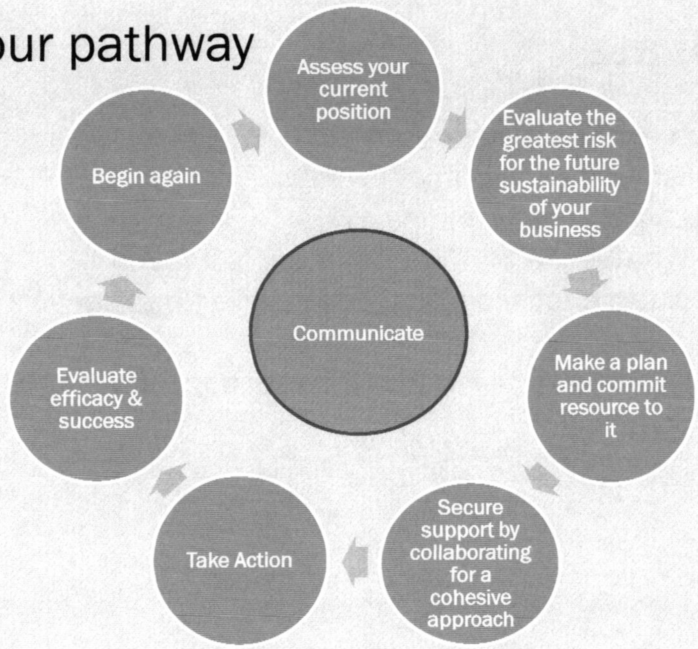

SUSTAINABLE LEATHER FOUNDATION

Assess

- Be honest
- Assess both internal and external factors
- Engage whole company in process
- Consider all 3 pillars:
    - How will this impact affect the environment?
    - How will this impact affect people and local economies?
    - How will this impact affect the financial success of the business?

Before you know where you can go, you first need to know where you are

Evaluate

- What is the biggest priority?
- Consider the following:
    - Largest positive impact?
    - What research / development is needed?
    - Is there technology already available to support?
- Talk with your customers and your suppliers – their priorities could differ.

Plan

- Put the necessary resources in place:
  - Assign responsibility
  - Allocation sufficient resources
    - Financial
    - Manpower
    - Infrastructure
- Communicate your plan to ensure continued accountability

Engage Support

- Other similar industry organisations – same challenges – work in a pre-competitive way for strength of purpose
- Customers and Suppliers – build strong partnerships that will support the process:
  - Trials
  - Communication
  - Shared financial burden
- Industry Associations and Support Organisations
- Educational Institutes – harness the young talent coming through

Take Action

Listen and work together

Overcome the obstacles

DO THE WORK!

Harness the minds of our industry

Innovate from within

Results

- Critical Evaluation is necessary
- Create strong case studies
- Communicate successes
- Communicate the ongoing challenges to success
- Share the knowledge!

Start again

- Sustainability is an evolving and continuous discipline
- Innovation from within our industry is what will protect it for future generations
- Our industry is under threat because for too long we have struggled to fully harness the knowledge of our technologists, chemists and scientists

> "Let's be the change that secures the future"

SUSTAINABLE LEATHER • FOUNDATION

If our industry wants to be able to fully counter the rise of alternatives we must be able to provide answers:

Measurable objectives
Verifiable data
Agreed parameters and principles

Stop talking and start walking! The pathway is there!

SUSTAINABLE LEATHER • FOUNDATION

"If we want our leather industry to be sustainable for future generations we have to act now!"

Thank you!

More information or further discussion:
Deborah@Sustainableleatherfoundation.com

Design Guidelines and Canvas for More Sustainable Leather Products – The Role of Design as a Driver for Sustainable Leather Goods in the 21st Century

Jonas Rehn[1]

Keywords: design guidelines, leather product design, design for sustainable development

1 Darmstadt University of Applied Sciences

jonas.rehn@h-da.de

sne.h-da.de/leather-chemistry

ABSTRACT:

Leather as an ancient material features characteristics and material properties that in many cases relate to sustainability by default (e.g. durability, repairability and aesthetically pleasing ageing). At the same time, in contrast, highly complex and non-transparent leather supply chains and increasing functional and aesthetic demands lead to problematic conditions regarding, for instance, the leather chemistry.

In line with regulatory developments (e.g. the EU Green Deal and related policies) and growing consumer awareness, many companies in the leather supply chains are striving to use "more sustainable chemistry" and to produce "more sustainable" leather goods. In the early phase of product development, design decisions (including sourcing and LCA) offer big opportunities in this competitive market. However, because of the technical and structural complexities, it is hardly possible to provide a list of design aspects that lead to more sustainable leather goods by default. The herewith presented concept of "Leather Design Guidelines for Sustainable Development", including the "Leather Product Design Canvas", aims to support the product development process of leather goods in favour of sustainable development by offering methods, case studies and paradigms that help designers, product managers and other parties involved in this process.

Based on the results of a transdisciplinary research project (including a scenario process) both guidelines and canvas as part these guidelines are being developed together with representatives from organisations along the leather value chains, as well as NGOs, as part of the project "Systeminnovation: Nachhaltige Entwicklung (s:ne)" at the Darmstadt University of Applied Sciences. This project relates directly to three other subprojects on traceability of chemicals along the supply chains, on chemical and process innovations for a more sustainable leather chemistry and, building on the results from the other subprojects, on harmonisation of standards.

This paper outlines the work done so far in drafting these guidelines. In order to present the overarching framework for the guidelines, the "Leather Product Design Canvas" draws links between general design process aspects and approaches for sustainable development that relate specifically to leather products. From this perspective, innovative systemic solutions as well as

business models (i.e. product-as-a-service offerings) can be developed that go beyond current approaches of leather product design. In a later phase of this project, the guidelines will be applied to one or more products as pilot studies to evaluate them regarding usability and effectiveness.

1 Aiming for More Sustainable Leather Products

Sustainable development, as operationalised by the United Nations' Agenda 2030[2], is a complex and multi-faceted issue. Many of the challenges associated with sustainable development are often referred to as a (super) wicked problem[3]. As outlined by Levin et al.[4], super wicked problems do not only consist of a number of complex and interdependent challenges. Moreover, solutions to one problem might create another problem.

The same applies to the leather industry and leather chemistry in general as well as to the design of more sustainable leather products. The global leather supply chains are complex, diffuse and interdependent systems. Many different production steps across continents, carried out by different companies, are required before a consumer product reaches the place where it is sold. As with most other industries, throughout this process, enormous amounts of chemicals are applied, which can be harmful to humans and the environment, energy is used and CO2 is emitted. For a leather product to be declared "sustainable" to some extent, a systemic perspective is required.

This applies, in particular, with regards to the 11 targets summarised by SDG 12 "Ensure sustainable consumption and production patterns". Designing more sustainable leather products involves a number of aspects, approaches and perspectives that relate to various aspects of sustainable development. Designers and others involved in the development process of new leather products often face the difficulty gaining an oversight of these issues and developing a systems perspective to create more sustainable products. While not all of these topics necessarily need to apply to each individual project or product, gaining an oversight and considering their role in the design process is

---

2 United Nations (2015): Transforming our World. The 2030 Agenda for Sustainable Development. Hg. v. United Nations.

3 Rittel, H. W. J.; Webber, M. M. (1973): Dilemmas in a general theory of planning. In: *Policy Sciences* 4 (2), S. 155-169.

4 Levin, Kelly; Cashore, Benjamin; Bernstein, Steven; Auld, Graeme (2012): Overcoming the tragedy of super wicked problems: constraining our future selves to ameliorate global climate change. In: *Policy Sci* 45 (2), S. 123-152. DOI: 10.1007/s11077-012-9151-0.

crucial to making an informed design decision. This requires an interdisciplinary approach that avoids "silo thinking", simplistic solutions and rebound effects (negative side effects caused by a singular improvement).

As a means to quickly cover all these topics in a first easy-to-access manner, an interdisciplinary group of experts, including, inter alia, design researchers, chemists, leather experts, developed the following Leather Product Design Canvas. It is based on the concept of the „Business Model Canvas" (BMC) proposed by Osterwalder in 2005 and refined since then[5]. According to van Boeijen and colleagues the canvas „as a tool can be used in various stages of the development process [...] The Business Model Canvas serves as a checklist to generate business ideas; it also structures, discusses, and evaluates these ideas on a conceptual level"[6].

The Leather Product Design Canvas[7] makes use of a similarly conceptual and graphical structure while comprising more refined elements and specific impulse questions to raise awareness and guide design processes. It aims at supporting the product development and design process of leather products to create more sustainable products. This shall be achieved by providing a quick and multi-perspective overview of relevant topics and considerations. One reason for this approach is that in many cases the designers of leather products such as shoes or bags do not necessarily need to be leather experts. Even interdisciplinary design teams might not include experts that understand both issues of sustainable development and material properties of this versatile material. Moreover, the complexity of the material leather and its various types of application would usually require a

---

5 Osterwalder, Alexander; Pigneur, Yves (2013): Business model generation. A handbook for visionaries, game changers, and challengers. New York: Wiley&Sons.

6 van Boeijen, Annemiek; Daalhuizen, Jaap; Zijlstra, Jelle (2020): Delft design guide. Perspectives, models, approaches, methods. Revised edition. Amsterdam: BIS Publishers. p. 135.

7 Rehn (2021): Leather Product Design Canvas. Darmstadt University of Applied Sciences, Germany. DOI: 10.48444/h_docs-pub-253

number of different experts to cover the most relevant topics related to sustainable development in order to identify possible rebound effects and systemic interdependencies of certain design decisions. The herewith presented canvas does by no means compensate for these expertise but aims at empowering design teams to take first steps towards more sustainable leather products based on a broadened scope while identifying knowledge gaps or the need for specific experts to be consulted or included in the design team.

One of the key challenges in designing more sustainable products is to improve specific aspects of a product or its ecosystem while avoiding purely "symbolic" actions, managing potential target conflicts and without creating negative side effects regarding another aspect. Therefore, this canvas illustrates relevant elements at one glance, raising awareness for their interdependencies. The Canvas, however, does not replace thorough analysis of these aspects, but rather aims at providing a multi-perspective overview and explorative impulse to foster more sustainable approaches. In line with this, the canvas is not a scientific analysis tool that results in more sustainable leather products by default. Its main function is to illustrate a comprehensive overview that informs and guides the design process through structure and impulse questions. In this way, otherwise potentially neglected issues such as pre- or post-consumer recyclability or sourcing is put into the centre of awareness.

1.1 Background and Development of this Canvas

This canvas is a first spin-off from the project "design guidelines for more sustainable leather products"[8] as part of a transdisciplinary research project at Darmstadt University of Applied Sciences aiming for more sustainable chemistry in the leather supply chains[9]. It is part of a comprehensive transformative research project aiming for system innovation for sustainable development funded by

8 to be published by Darmstadt University of Applied Sciences in 2022

9 System innovation for more sustainable leather chemistry: https://sne.h-da.de/en/implementation-project/more-sustainable-chemistry-in-the-leather-supply-chains (last checked on October 19th 2021)

the Federal Ministry of Education and Research - BMBF and the State of Hesse (First funding period until 31.12.2022)[10]. The term "system innovation" refers to technical, social and organisational innovation, underpinned by an institutional framework[11], that is necessary to achieve transformation towards sustainable development.

Both the overall project and its "subprojects" are carried out by transdisciplinary teams including experts such as scientists from different fields (Law, Business, Psychology, Design, Chemistry and Engineering), representatives along the leather supply chains (chemicals, tanning companies, processing industry, brands, etc.) and other stakeholders such as NGOs and political consultants. Since the year 2018 the project team conducted a comprehensive participatory process that includes a six-month scenario process in which representatives from the leather supply chains co-created scenarios for the global leather industry for the year 2035. Based on this process a "theory of change" has been developed that aims at structuring a multi-stakeholder process towards a more sustainable leather chemistry by 2035[12]. In September 2020, the project started four multi-stakeholder implementation projects aiming at specific topics outlined in the theory of change that promise the greatest leverage in terms of the envisaged transformation towards more sustainable chemistry[13]. These subprojects focus on (a.) IT tools and governance for traceability of chemicals along supply chains, (b.) chemical and process innovations, (c) design guidelines for more sustainable leather products. and (d.) harmonisation of standards for "more sustainable leather chemistry".

10 System innovation for sustainable development: https://sne.h-da.de/en/ (last checked on October 19th 2021)

11 Institution, in this context, refers to all formal and informal instructions such as legal rules, company policies, habitual patterns of designers etc., see North 1991.

12 Schenten, Julian; Rehn, Jonas (2021): A Theory of Change (ToC) supporting the visioning of a sustainable supply chain. Beitrag im Portal "SCNAT wissen - Methoden und Werkzeuge für die Koproduktion von Wissen". Akademien der Wissenschaften Schweiz Netzwerk für transdisziplinäre Forschung (td-net). https://naturwissenschaften.ch/co-producing-knowledge-explained/practical_experiences/theory_of_change, zuletzt geprüft am 22.03.2021.

13 For a recording of the kick-off-event visit: https://youtu.be/_ha6VG4NXUE

1.2 General Structure of the Canvas

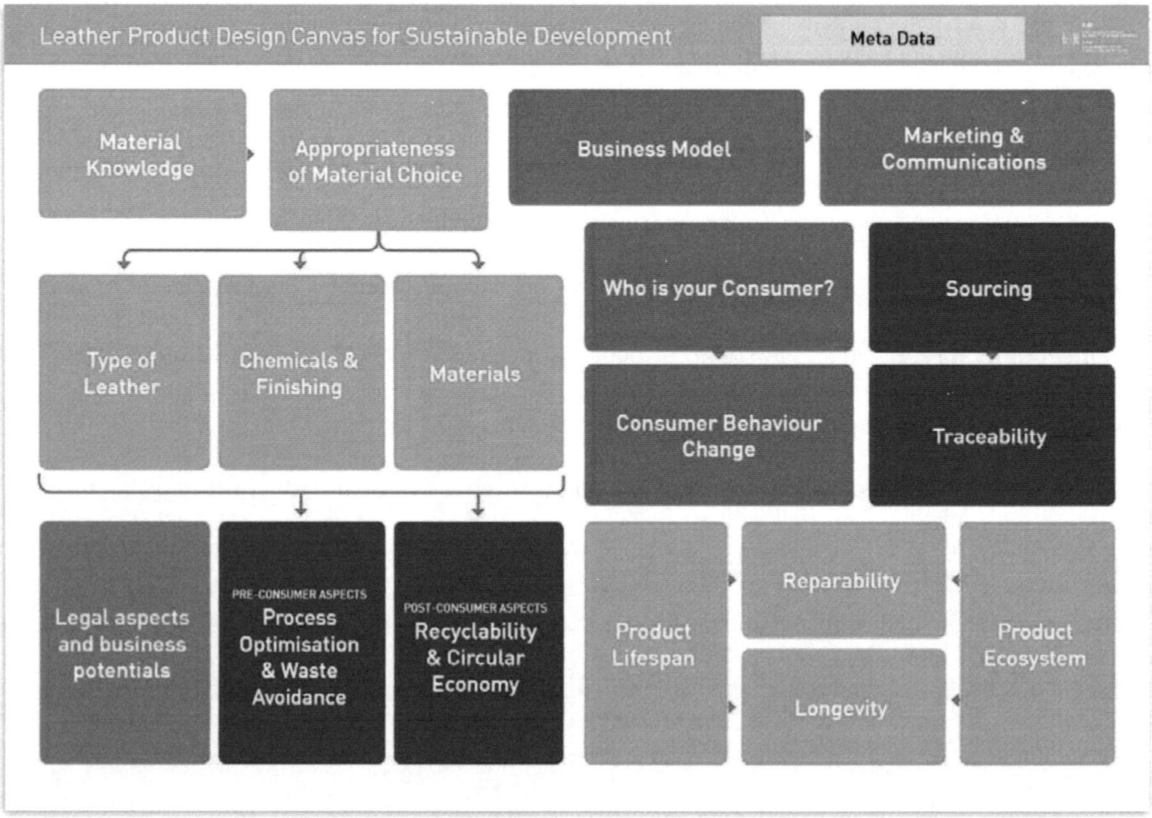

Fig. 1: Overall layout of the Leather Product Design Canvas

The Canvas covers four overarching themes divided into 18 boxes each resembling one relevant aspect of the leather product at hand with regards to sustainable development (Fig. 1). The green boxes in Fig. 1 refer to aspects that could be summarised as rather technical such as material properties, chemicals used or the technical aspects that allow or inhibit reparability. The brown boxes focus on procedural subjects. Although they as well include technical topics such as traceability, they primarily refer to questions or issues that affect the organisation of producing, transporting and disposing or recycling of the products. The purple boxes refer to economical and consumer topics including business models and the role, awareness and behaviour of consumers. The overarching legal and institutional framework that applies to the product at hand is represented by the orange box. It is closely linked to other aspects such as business models that are based on current regulatory developments.

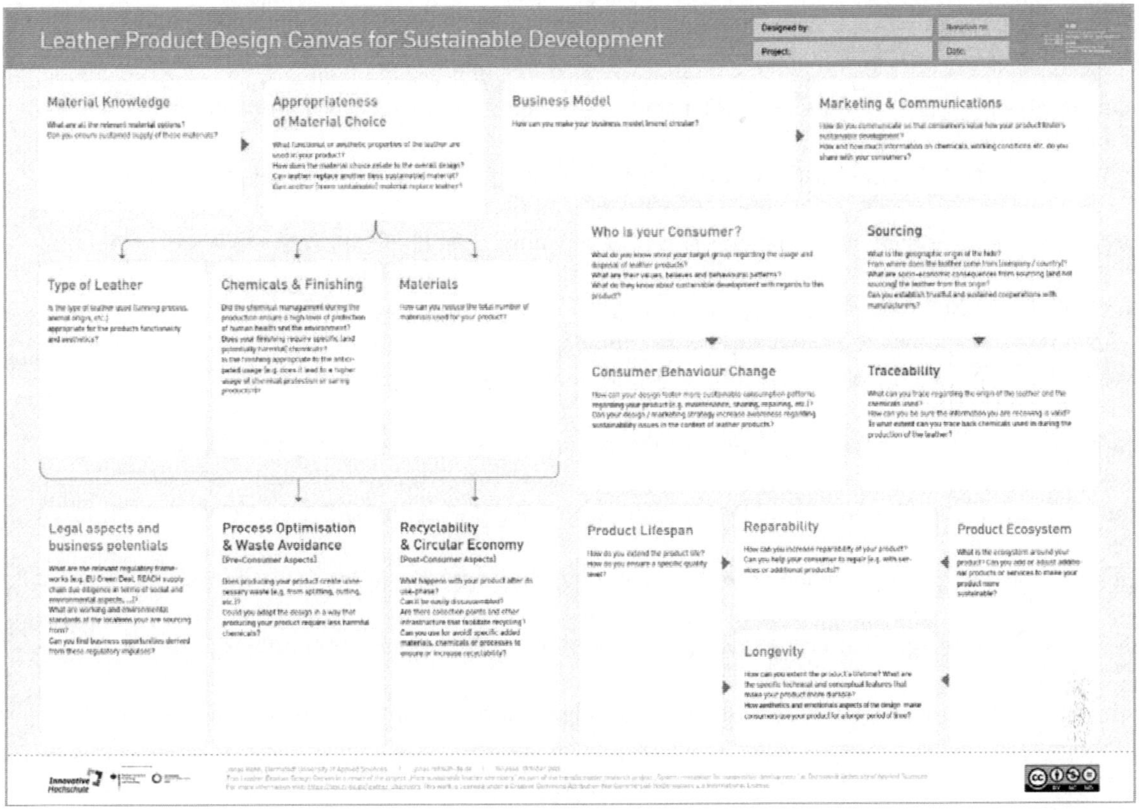

Fig. 2: Leather Product Design Canvas, DinA3 (Rehn, 2021)

In each of the 18 boxes, impulse questions specify the scope of these categories (Fig. 2). These questions are meant to guide the design process and initiate considerations to improve the design concept. As each project, product and supply chain is different, the size of boxes or number of impulse questions do not reflect the importance or relevance of a specific topic.

Additional arrows and lines indicate clusters, interdependencies and direct relations. However, as leather supply chains are complex systems, all boxes are related to each other to some extent. Therefore, these indications should be seen as rather most basic or obvious relations and clusters.

1.3 Users of the Canvas

The complexity inherent to the field of sustainable development requires an interdisciplinary systems thinking approach. With regards to leather products, designers might be faced with contradicting targets that make it even more difficult to define a proper briefing for more sustainable products. The fashion industry, for instance, is demanding for continuously new styles and approaches that often follow a business model of planned obsolescence. Instead, more sustainable and innovative leather products (and related services for that matter) could include business models or strategies that have a profound impact on many other departments of a company than product development such as sourcing, quality management, business development, marketing etc. Understanding the paramount role design might play to create more sustainable products and influence more sustainable consumer behaviours requires a paradigm shift that puts design decisions at the core of strategic management. Therefore, design cannot be done by one person alone but rather needs to be a team effort across disciplines.

This being said, ideally, the canvas shall be used by an interdisciplinary team comprising as many roles as possible that relate to the design and development of new leather products (such as sourcing, marketing, etc.). As this mode of work poses additional challenges and threats for a design process, it is recommended to use well-prepared workshops when applying this canvas in interdisciplinary teams. In this way, the designer's role might be extended from simply designing products to actually facilitating and moderating the design process, taking all relevant impulses and requirements from other disciplines into account. At the same time, to get the most out of this canvas to thoroughly design leather product while considering as many relevant sustainability issues as possible might require a change of corporate culture or mind-set in some organisation, fostering open-minded and transparent communication across departments and disciplines.

1.4 Mode of Work

As mentioned before, the canvas is not meant to be a strict formula or checklist but rather an exploratory analysis and conceptualisation tool that can be used and revisited (iterative process) at any stage throughout the design process. Boxes and questions do not necessarily have to be addressed in their order of appearance.

While not all impulse questions might apply to each project, they could raise awareness or identify gaps of knowledge that need to be addressed before moving on in the design process. Therefore, the canvas is both explorative inspiration as well as a simplified evaluation tool.

The canvas can be fulfilled by one person alone or as part of a workshop or team effort. It is recommended to use the canvas in an iterative manner. Additional information boxes in the header of the canvas support this iterative mode of work. As illustrated in fig. 3, it might be useful to note down ideas and concepts directly on the canvas. This also encourages short and pointed descriptions due to limited space available. At the same time, the canvas can be used and adjusted to the needs of each individual project. Therefore, team workshops might require more space or additional worksheets to go deeper into specific boxes or topics.

Especially with regards to workshops, it can be useful to combine the canvas with other tools and techniques from the field of design thinking[14]. Workshops applying this canvas should be planned and orchestrated according to the usual principle of design- and brainstorming workshops, curating an open mind-set and appreciating new, disruptive and sometimes unrealistic ideas to go beyond

[14] Kumar, Vijay (2013): 101 design methods. A structured approach for driving innovation in your organization. Hoboken, N.J: Wiley.

current concepts and solutions. After each workshop, the canvas worksheets should be analysed in detail by the workshop facilitator to extract insights and concepts that otherwise might be lost.

One of the most important aspects of working with the canvas is to identify and indicate relations and interdependencies. Therefore, it is recommended to use arrows, lines and other graphical or written information to illustrate connections and foster systems thinking (see e.g. fig. 3).

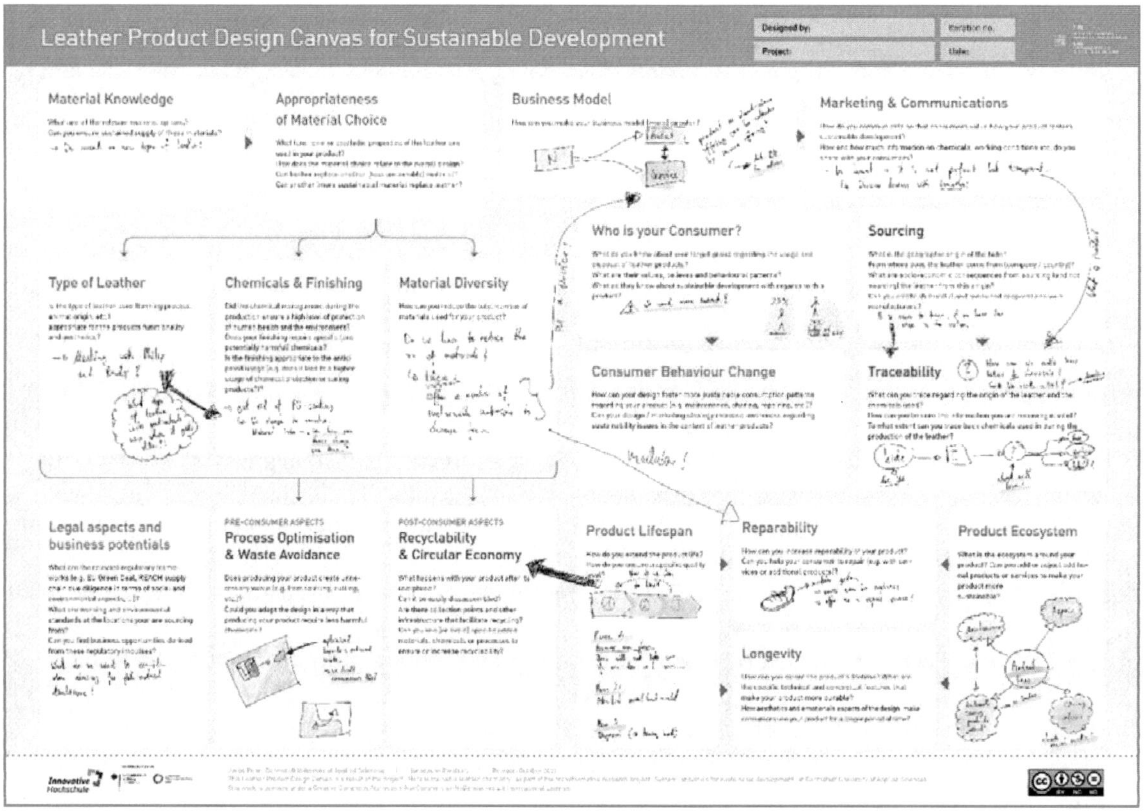

Fig. 3: Applied Leather Product Design Canvas for female dress shoe

It is recommended to use the Leather Product Design Canvas while applying the aforementioned Design Guidelines for Sustainable Development. While the guidelines aim at deepening one's knowledge and broadening one's scope with regards to more sustainable leather products, they shall be seen as both piece of reference and exploratory source of inspiration like comparable guidelines in the field of circular economy[15] or even public design[16].

1.5 Open-End Template

Although this canvas has been co-developed and tested with industry partners and an interdisciplinary group of scientific experts, there might be projects or phases for which important aspects or new trends are entirely missing. Users of the canvas should therefore feel free to adjust and extent the canvas according to the needs of the project. We encourage users who have feedback and ideas for improvement to get in touch with us.

---

15 e.g. Nike, Inc. (Hg.) (2021): Circularity Workbook. Guiding the Future of Design. Online verfügbar unter https://www.nikecirculardesign.com/guides/CircularityGuide.pdf, zuletzt geprüft am 16.03.2021.

16 e.g. Center for Active Design (2010): Active Design Guidelines. Promoting physical activity and health in design. Hg. v. City of New York. New York. Online verfügbar unter https://centerforactivedesign.org/guidelines/, zuletzt geprüft am 03.12.2016.

Literature

van Boeijen, Annemiek; Daalhuizen, Jaap; Zijlstra, Jelle (2020): Delft design guide. Perspectives, models, approaches, methods. Revised edition. Amsterdam: BIS Publishers. p. 135.

Center for Active Design (2010): Active Design Guidelines. Promoting physical activity and health in design. Ed. by City of New York. New York. Online verfügbar unter https://centerforactivedesign.org/guidelines/, last checked 03.12.2016.

Kumar, Vijay (2013): 101 design methods. A structured approach for driving innovation in your organization. Hoboken, N.J: Wiley.

Levin, Kelly; Cashore, Benjamin; Bernstein, Steven; Auld, Graeme (2012): Overcoming the tragedy of super wicked problems: constraining our future selves to ameliorate global climate change. In: *Policy Sci* 45 (2), S. 123-152. DOI: 10.1007/s11077-012-9151-0.

Nike, Inc. (Hg.) (2021): Circularity Workbook. Guiding the Future of Design. Online verfügbar unter https://www.nikecirculardesign.com/guides/CircularityGuide.pdf, last checked 16.03.2021.

North, Douglass C. 1991. Institutions. *The Journal of Economic Perspectives* 5: 97–112.

Osterwalder, Alexander; Pigneur, Yves (2013): Business model generation. A handbook for visionaries, game changers, and challengers. New York: Wiley&Sons.

Rehn (2021): Leather Product Design Canvas. Darmstadt University of Applied Sciences, Germany. DOI: 10.48444/h_docs-pub-253.

Rittel, H. W. J.; Webber, M. M. (1973): Dilemmas in a general theory of planning. In: *Policy Sciences* 4 (2), p. 155-169.

Schenten, Julian; Rehn, Jonas (2021): A Theory of Change (ToC) supporting the visioning of a sustainable supply chain. Beitrag im Portal "SCNAT wissen - Methoden und Werkzeuge für die Koproduktion von Wissen". Akademien der Wissenschaften Schweiz Netzwerk für transdisziplinäre Forschung (td-net). https://naturwissenschaften.ch/co-producing-knowledge-explained/practical_experiences/theory_of_change, last checked 22.03.2021.

United Nations (2015): Transforming our World. The 2030 Agenda for Sustainable Development. Ed. by United Nations.

Appendix

Leather Product Design Canvas – A3-template (PDF)

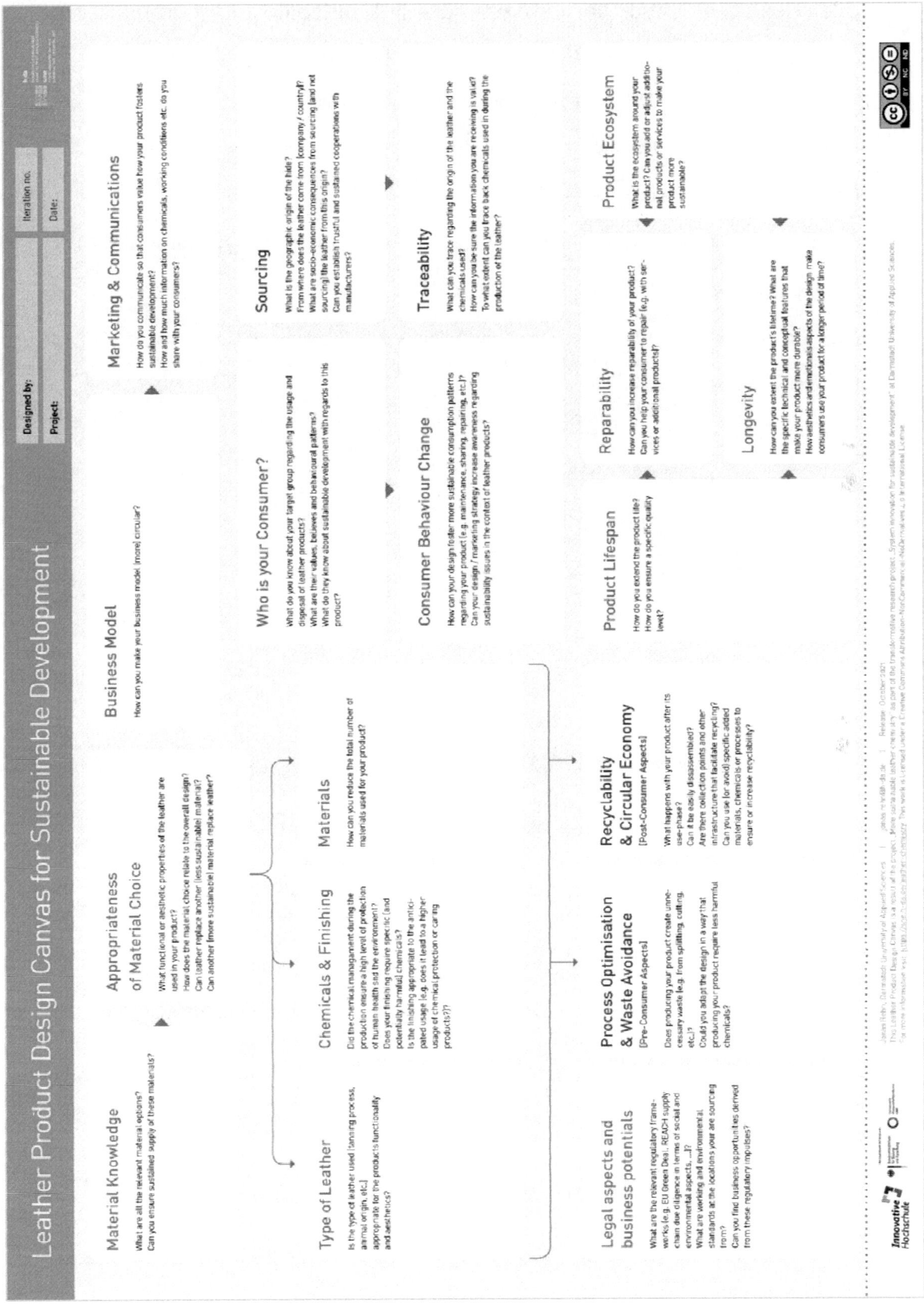

An antibacterial strategy on chrome-free tanned leather: based on β-cyclodextrin aldehyde derivatives and small molecule antimicrobial agent

Wen Huitao[1,2], Yu Guofei[1,3], Ding Zhuang[1,3]

Chen Yining[1,3], Dan Nianhua[1,3], Dan Weihua[1,2]*

[1]National Engineering Research Center of Clean Technology in Leather Industry, Sichuan University, Chengdu 610065, China

[2]Fujian Key Laboratory of Green Design and Manufacture of Leather, Xingye Leather Technology Co., Ltd., Jinjiang 362271, China;

[3]Key Laboratory of Leather Chemistry and Engineering of Ministry of Education, Sichuan University, Chengdu 610065, China

*Corresponding authors.

E-mail: danweihua_scu@126.com

Fax: (+86)-028-85468589; Tell: (+86)-028-85468589

Abstract

Chrome tanning has dominated in leather production up to now, while facing a significant challenge in terms of chromium. Therefore, the chrome-free tanning process becomes an inevitable choice. However, chrome-free tanned leather (CFTL) is poor in antibacterial properties, which limits its practical application in leather production. In this work, a new environment-friendly antibacterial strategy based β-cyclodextrin (β-CD) aldehyde derivatives and ciprofloxacin hydrochloride (CH) as a small molecule model antibacterial agent was developed to address this issue along with improving the comprehensive performances of leather. The structure and properties of obtained leather were characterized by Fourier transform infrared spectroscopy, scanning electron microscopy, differential scanning calorimetry, thermogravimetric, and its antibacterial effect against *E. coli* and *S. aureus* was investigated. The results showed that compared with CFTL without *β*-CD modification and CH treatment, the T_s, T_d and T_p of the leather with this method possessed increased 5.7°C, 4.9°C, and 6.9°C, respectively. After being treated by CH with 90 days of storage, the antibacterial rates against *E. coli* and *S. aureus* still reached 99.8% and 97.4%, respectively, which were 29.9% and 34.4% higher than those without *β*-CD treatment. It, that antibacterial agent is included in the cavity of *β*-CD on CFTL, will improve the poor antibacterial durability due to the release of the small molecule antimicrobial agent. These findings, therefore, indicated that a new antibacterial method on CFTL with *β*-cyclodextrin aldehyde derivatives and ciprofloxacin hydrochloride has potential practical application prospects in leather production.

Keywords: chrome-free tanned leather, antibacterial performance, *β*-cyclodextrin, ciprofloxacin hydrochloride

1. Introduction

Leather is a natural material used by mankind for a long time, and it plays an important role in daily life, economy, military, and other aspects (Sathish *et al.*, 2016). Up to now, Chrome tanning has dominated in leather production, while facing a significant challenge in terms of chromium (Kanagaraj *et al.*, 2020). Therefore, the chrome-free tanning process, such as tannin extract, organic tanning agent and metal complex tanning agent, had become an inevitable choice (Krishnamoorthy *et al.*, 2012). However, chrome-free tanned leather (CFTL) is poor in antibacterial properties (Orlita, 2004), which limits its practical application in leather production.

The conventional solution was to add antibacterial agents during the tanning process, and the antibacterial agents were mainly distributed on the surface of the leather and between the collagen fibers. It should be noted that to reduce the adverse impact of antibacterial agents on leather performance and the environment, the addition amount should be little (only 0.1-0.3% of the weight of wet leather), so that the antibacterial performance was poor and difficult to last. Related researches tend to choose more efficient and safe fungicides such as nanosilver (Velmurugan *et al.*, 2014), chitosan (Barros *et al.*, 2011), quaternary ammonium salts (Xiao et al., 2014), compound fungicides (Cuadros *et al.*, 2012), and encapsulated fungicides (Geng *et al.*, 2018), but the effects were not ideal, and the problem has not been completely solved so far.

Herein a strategy is to introduce a complete *β*-CD cavity on CFTL, and fix the small molecule antibacterial agent in the cavity, and use its slow-release effect to achieve the purpose of long-term antibacterial. β-CD, a natural product of CGTase acted on starch (Valle and Martin, 2004), was a cyclic oligosaccharide consisting of seven glucopyranose units linked by α-(1,4) bonds, had a hydrophilic outer surface and a hollow hydrophobic interior cavity. The most notable feature was that it could form stable inclusion complexes (host-guest complexes) with a wide variety of hydrophobic guest molecules in the cavity (Eastburn and Tao, 1994; Radu *et al.*, 2016; Matencio *et al.*, 2020). Inclusion complexes had a significant influence on the physicochemical properties of guest molecules, and the guest molecules were temporarily locked or caged within the host cavity, such as improved stability, reduced volatility, and sublimation, as well as controlled and sustained release.

To enhance the force between β-CD and leather, β-CD was oxidized into β-CD aldehyde derivatives (CDAD). Meanwhile, for achieving the effect of multi-point cross-linking, each molecule of CDAD should have more than two aldehyde groups. Different from the commonly used 2, 3-dialdehyde β-CD, CDAD used in this work is multiple-6-deoxy-6-formyl-β-CD, prepared by oxidizing the primary hydroxyl group on the C_6 position, without destroying the cavity structure of β-CD (Wen *et al.*, 2021).

Ciprofloxacin hydrochloride (CH), a quinolone carboxylic acid derivative, could be used as a model antibacterial agent to apply in textile antibacterial materials (Hou *et al.*, 2021). β-CD and CH could form a 1:1 inclusion compound (Chao *et al.*, 2002). After the inclusion, CH could have a slow-release effect, which made the antibacterial performance more stable and lasting (Choi *et al.*, 2017).

In this work, CDAD modified CTFL (CD-CFTL) was prepared by Schiff base reaction, and CH antibacterial treated CFTL (CH/CD-CFTL) was prepared by inclusion reaction of *β*-CD on CD-CFTL (Fig. 1). The structure and properties of as-formed leather was characterized by Fourier transform infrared spectroscopy (FTIR), scanning electron microscopy (SEM), differential scanning calorimetry (DSC), thermogravimetric (TGA), and its antibacterial effect against *E. coli* and *S. aureus* was investigated. The results show that modified CFTL had improved thermal properties and excellent antibacterial properties, which not only provides a new way to improve the antibacterial properties, but also provides a new idea to prepare

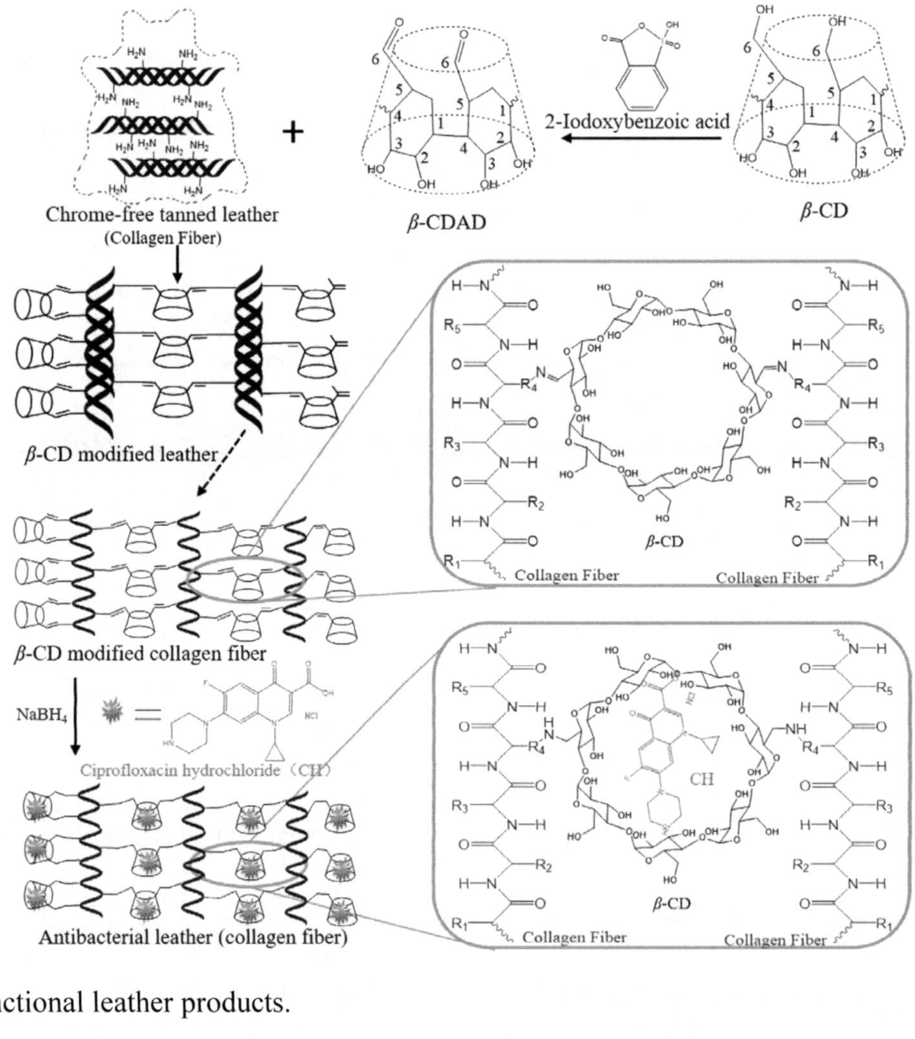

multifunctional leather products.

Fig.1 Preparation of antibacterial CFTL

2. Experimental

2.1 Materials

CH, for research use only, Beijing Solarbio Science & Technology Co., Ltd. *β*-CD was a biological reagent and purchased from Chengdu Kelong Chemical Co., Ltd (China). *β*-CD aldehyde derivatives (*β*-CDAD) was prepared according to an improved method (Hu *et al.*, 1999), the content of aldehyde groups of the per *β*-CD molecule was 2.2, measured by hydroxylamine hydrochloride-potentiometric titration method (Zhao and Heindel, 1991). All the other reagents used were of analytical grade and purchased from Chengdu Kelong Chemical Co., Ltd (China).

The chrome-free tanned leather (CFTL) was prepared by the combination tannage (Wen et al., 2021) based on an organic tanning agent F-90 (Granofin® Easy F-90), tannin extract and Zr-Al-Ti complex tanning agent (CAT), which has a tanning effect comparable to that of chrome tanning, and the tanning process was shown in Tab. 1.

Tab.1 The tanning process of the chrome-free tanned leather

Process	Chemicals	Dosage /%	Temp. /°C	time / min	pH
Washing	H_2O	250	25	10	7.0
Tanning I	H_2O	30	30		
	F-90	5.0		120-180	
	H_2O	30	38-42	120	5.5
Tanning II	HCOOH(1:10)	0.5	25	30	4.0-4.5
	sulfite fish oil	1		30	
	tannin extract	15		180	drain
Tanning III	H_2O	50	35	120	
	HCOOH(1:10)	0.5	25	30	2.8-3.0
	HCOONa	1.0			
	CAT	6.0		180	
	$NaHCO_3$	2.0			
	H_2O	150	40	180	3.8-4.0
Washing	H_2O	150		10	Drain

2.2 Samples preparation

2.2.1 *β*-CDAD modified CFTL

First, the CFTL was put into a drum with a speed of 6-8 r/min, containing 50-80% of 30-32°C water, and the pH was adjusted to 2.5-4.5, rotated for 30 minutes. Then *β*-CDAD (3-7%, based on the weight of CFTL) was added, reacted for 2-6 hours. And then the pH was adjusted to 4.5-6.5, the temperature was adjusted to 38-42°C, and the reaction time was 2-6 hours. Finally, add a certain amount of sodium borohydride and continue the reaction for 2-6 hours. The *β*-CDAD modified CFTL (CD-CFTL) were made by natural drying after washing.

Theoretically, the weight gain rate of CFTL after *β*-CDAD modification was the *β*-CD graft rate, which was calculated according to formula (1):

$$W\% = \frac{m_1 - m_0}{m_0} \times 100\% \qquad (1)$$

where W represents the weight gain rate of CFTL after CDAD modification, %; m_1 was the weight of CFTL after *β*-CDAD modification, g; m_0 was the weight of CFTL, g.

2.2.2 CH antibacterial treated CD-CFTL

First, the CD-CFTL were put into a drum with a speed of 6-8 r/min, containing 200% of 25-45°C water, and the pH was adjusted to 4.0-6.0, rotated 30 minutes. Then CH (0.01-0.05%, based on the weight of CD-CFTL) was added and reacted for 2-10 h. Finally, the resultant antibacterial leather was washed twice with water, and then piled for further evaluation.

The included rate of CH and the CH loaded degree of CD-CTFL were calculated according to Eq. (2) and Eq. (3), respectively.

$$N = \frac{C_0 - C_1}{C_0} \times 100\% \quad (2)$$

$$L = \frac{N * C_0 * V}{m_{CTFL}} \quad (3)$$

where *N* represents the included rate of CH, %; *L* represents the CH load degree of CD-CTFL, mg/g; C_0 and C_1 is the concentration of CH at the start and certain time during the load process, mg/L, V is the volume of CH solution and m_{CTFL} is the weight of CTFL.

2.3 Characterization

2.3.1 FTIR spectra

The FTIR spectra were recorded using an FTIR spectrometer (Nicolet is10, Thermo Scientific, USA). Each about 1 mg sample was pressed into a thin sheet with 200 mg KBr pellet, and measured in the range of 400-4000 cm^{-1} at room temperature, using 32 scans and 4 cm^{-1} resolution under the conditions of a temperature of about 25°C and a humidity of about 65%.

2.3.2 UV-vis Spectra

The UV-vis spectra were recorded using a UV-vis spectrometer (U-3900H, HITACHI, Japan). After UV-vis full-wavelength scanning, in an acetic acid-sodium acetate buffer solution with a pH of about 4.0, CH has an absorption peak at 276 nm. The waste liquid after the CH antibacterial treatment was adjusted to pH 4.0, and the absorbance at 276 nm was measured.

2.3.3 SEM

The morphologies were observed using an SEM (Apreo S, Thermo Scientific, America). Samples were frozen section, removed the surface layer of equal thickness, fixed on the conductive adhesive, sputter and coated with gold, and then observed with an accelerated voltage of 15 kV.

2.3.4 Thermal properties

Shrink temperature (T_s) was tested using a digital leather shrinkage temperature tester (MSW-YD4, Shaanxi University of Science & Technology, China) in light of ISO 3380:2015. Thermal properties were recorded using a DSC analyzer (204 F1, Netzsch, German) and a TGA analyzer (209 F1, Netzsch, German).

2.3.5 Antibacterial performance test

The antibacterial performance test against *E. coli* and *S. aureus* were carried out refers to GB/T 21510-2008 Appendix II: Material antibacterial performance test method: oscillation method (Chao *et al.*, 2014). And the antibacterial rate was calculated according to Eq. (4):

$$A=\frac{N_0-N_1}{N_0}\times 100\% \quad (4)$$

where A represents the antibacterial rate, %; N_0 and N_1 were the number of viable bacteria in the blank group and the test group, respectively.

3. Results and discussion

3.1 Preparation and characterization of CD-CFTL

CD-CFTL was prepared by Schiff base reaction between aldehyde groups on β-CDAD and amino groups on CFTL (Bowes and Cater, 1968) . The important structural feature was the introduction of the β-CD cavity, while the content of amino groups decreased and the content of hydroxyl groups increased.

3.1.1 Experimental optimization

The experimental design was performed to evaluate the influence of the following variables on the modified process, to determine the best process conditions: the amount of β-CDAD, the pH of the permeation and binding, and the time of permeation, binding reaction, and reduction reaction.

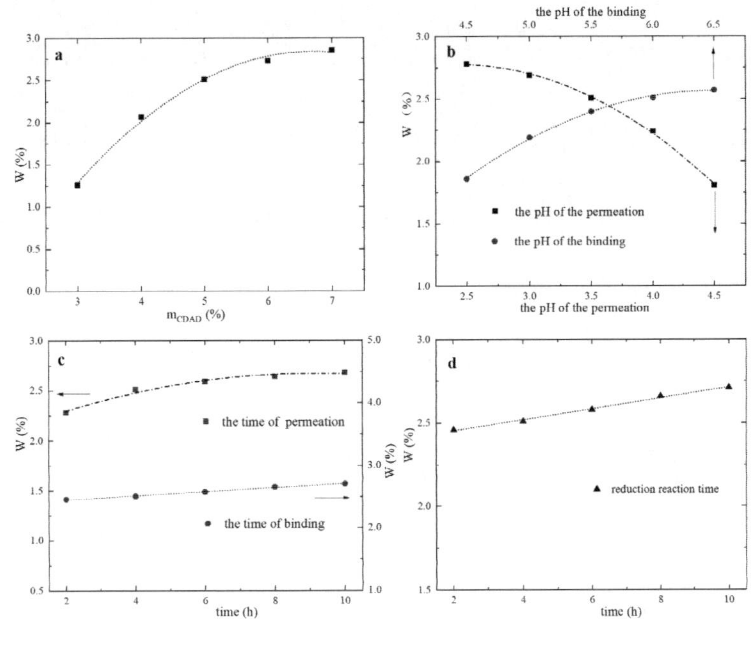

(a) the amount of CDAD, (b) the pH of permeation and binding reaction,

(c) the time of permeation, binding, and reduction reaction

Fig.2 Effect of reaction conditions on weight gain rate

Using a single factor experimental method and the weight gain rate as an evaluation index, the dosage of β-CDAD, pH and time of osmotic reaction, pH and time of binding reaction, and reduction reaction time were optimized one by one. The results (Fig.2) show when the amount of *β*-CDAD is 5%, the pH and time of the osmotic reaction are 3.5 and 4 hours, the pH and time of the binding reaction are 6.0 and 4 hours, and the reduction reaction time is 2 hours, an ideal weight gain effect can be achieved, with a weight gain rate about 2.51%.

3.1.2 FTIR analysis

CDAD is composed of multiple hydroxyl glucose units, the peak at 3402 cm^{-1} is O–H tensile vibration, the peak at 2925 cm^{-1} is C–H tensile vibration, the peak at 1610 cm^{-1}, 1160 cm^{-1}, 1075 cm^{-1}, 1030 cm^{-1}, 850 cm^{-1} are all characteristic peaks of D-glucopyranose. Since the 6-position primary hydroxyl group is partially oxidized to an aldehyde group, a characteristic peak of the aldehyde group appears at 1732 cm^{-1} (Chao *et al.*, 2014).

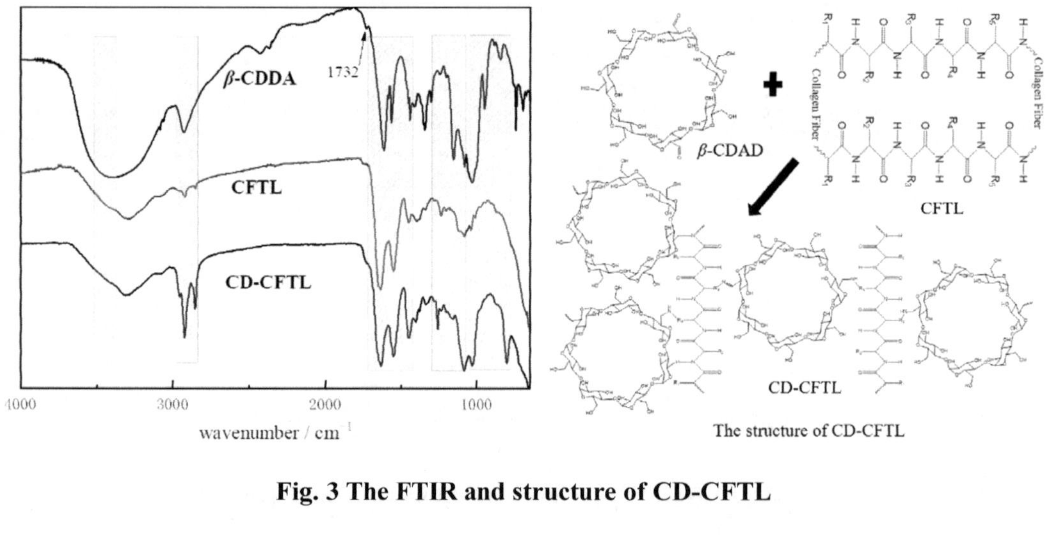

Fig. 3 The FTIR and structure of CD-CFTL

CFTL was tanned by combining F-90, tannin extract and Zr-Al-Ti complex tanning agent. F-90 could react with the carboxyl and the amino groups, the tannin extract could be combined with the peptide bonds and amino groups, and the Zr-Al-Ti complex tanning agent could react with the carboxyl group of CFTL, and the synergy effect between a variety of chromium-free composite tanning agents, giving the excellent tanning effect. The principle to prepare CD-CFTL was based on the Schiff base reaction between the aldehyde group of β-CDAD and the amino group of CFTL, which was similar to the aldehyde tanning of leather. Fig.3 showed the amide A peak was blue-shifted, the peak intensity of amides I, II, and III decreases, and the characteristic peaks of the β-CD structure appear. All these changes indicate that β-CD has been grafted onto CFTL (Wen *et al.*, 2021).

3.1.3 SEM analysis

The SEM of CFTL and CD-CFTL were showed in Fig. 5, which was removed the surface of equal thickness by a freezing section. The collagen fiber bundles of CFTL were relatively regular and arranged relatively tightly (Fig. 5a). After β-CDAD modification, many dots appeared on the SEM of CD-CTFL. These dots were considered to be the grafted β-CD, which was closer to the SEM of β-CD grafted wool (Yu *et al.*, 2016).

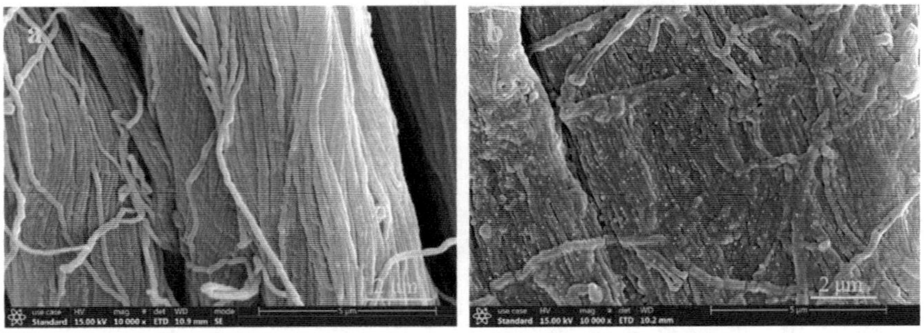

Fig.5 SEM of CFTL and CD-CFTL

3.1.3 Thermal analysis

The thermal shrink temperature (T_s), denaturation temperature (T_d), and decomposition temperature (T_p) of CFTL and CD-CFTL were shown in Fig. 6. After modification by CDAD, the T_s, T_d and T_p of CFTL increased 5.7 °C, 4.9 °C, and 6.9 °C, respectively, which showed CDAD could effectively crosslink the collagen fibers of CFTL and has a better tanning effect.

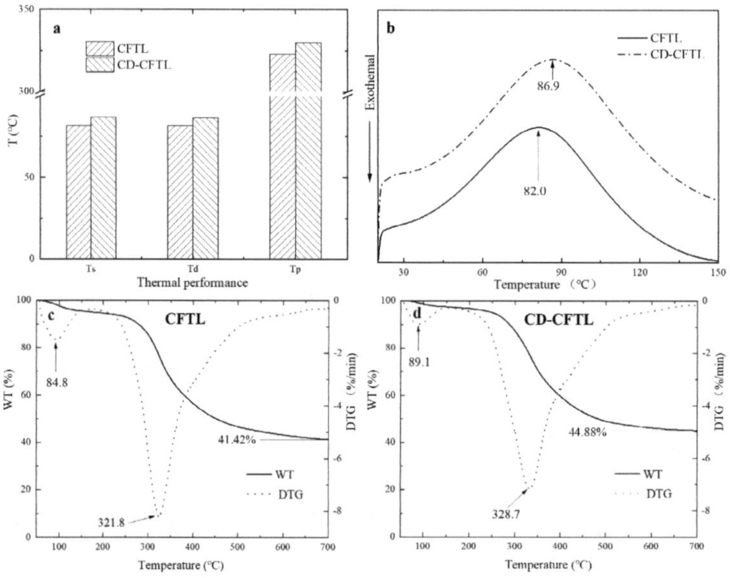

Fig. 6 The T_s, T_d, and T_p of CFTL and CD-CFTL

3.2 Preparation and characterization of CH/CD-CFTL

The characterization of CD-CFTL proved that *β*-CD has been grafted onto CFTL and maintains its original three-dimensional cavity, so it has a inclusion ability for small molecules. As far as *β*-CD is concerned, it is very suitable for the inclusion of small molecules with the five-membered ring, benzene, or naphthalene structures, such as imidazole, thiazoles, phenols, fluoroquinolones, vanillin, etc. Among these organic antibacterial agents, CH has good antibacterial activity against both gram-positive bacteria and gram-negative bacteria. It has the advantages of a wide antibacterial spectrum, strong antibacterial activity, low dosage, and small toxic and side effects, and can form a 1:1 inclusion compound with β-CD (Gauzit *et al.*, 2020), so CH is selected as a model antibacterial agent.

3.2.1 Experimental optimization

The experimental design was performed to evaluate the influence of the following variables on the antibacterial treatment process, to determine the best process conditions: the amount of CH, the pH, temperature (T), and time (t) of the antibacterial treatment process.

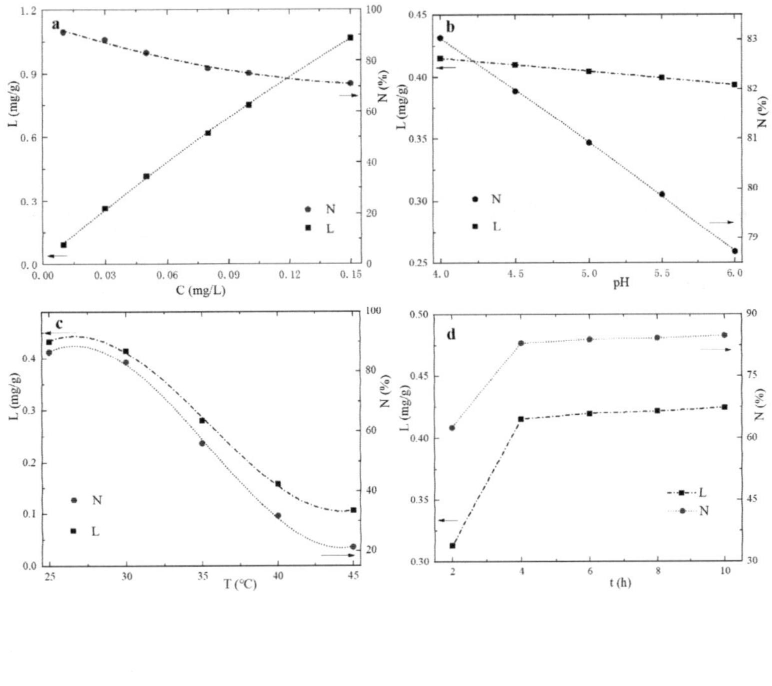

Fig.7 Experimental optimization of the antibacterial treatment process

Using a single factor experimental method, the included rate of CH and the CH loaded degree as evaluation indexes, the dosage of CH, pH, T, and t of antibacterial treatment process were optimized. The results (Fig.7) show that when the concentration of CH is 0.05 mg/L, the pH is 4.0, the T is 30°C, and time is 4 hours, an ideal included effect can be achieved, with the included rate of CH is 83.04% and the CH loaded degree of CD-CFTL is 0.42 mg/g.

3.2.2 FTIR analysis

Fig. 8 showed the FTIR and structure of CH, CFTL, CD-CFTL, CH/CFTL, and CH/CD-CFTL. The characteristic peaks of CH could be observed in CH/CFTL, but disappear in CH/CD-CFTL. The reason was that CH was included in the cavity of *β*-CD. Compared with the FTIR spectrum of CD-CFTL, the intensity of the amide I and amide II peaks on the CH/CD-CFTL spectrum had decreased and blue-shifted, amide A had a redshift, and the intensity of the CO vibration peak attributable to the D-glucopyranose unit of *β*-CD had decreased (Yap *et al.*, 2005). It could be found that the intensity of all peaks on CH/CD-CFTL has been weakened to varying degrees, which was caused by the obstruction of the molecular vibration after inclusion (Liu, 2008). All phenomena indicated that CH was included in the cavity of *β*-CD on CD-CFTL.

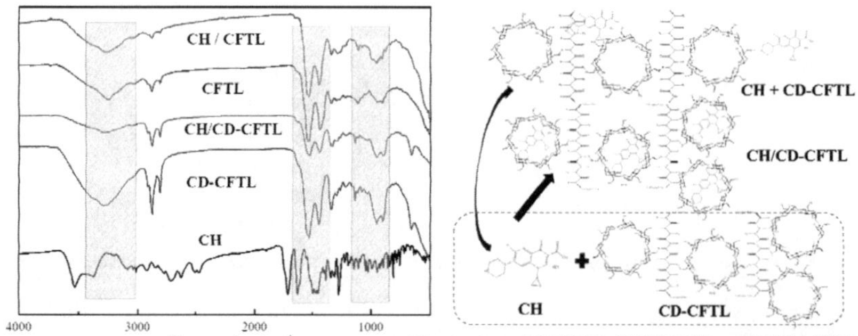

The structure of CH, CFTL, CH/CD-CFTL and CH+CD-CFTL

Fig. 8 The FTIR and structure of CH, CFTL, CD-CFTL, CH/ CFTL and CH/CD-CFTL

3.2.3 Thermal analysis

The T_s, T_d, and T_p of CH/CFTL and CH/CD-CFTL were shown in Fig. 9. Compared with CFTL, CD-CFTL after CH antibacterial treatment had better thermal properties. The reason was that CH and CD could form an inclusion compound, and the thermal stability of the inclusion compound was stronger (Zhang *et al.*, 2012). After CH was loaded, an exothermic peak appears at 157.4°C in CH/CFTL, which was confirmed to be a characteristic peak of CH (Zhu, 2008). And the peak did not exist in CH/CD-CFTL, which also showed that CH was included in the cavity of β-CD.

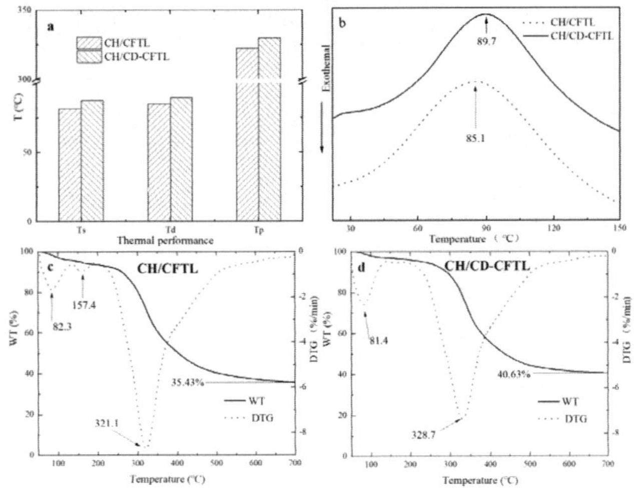

Fig.9 The T_s, T_d, and T_p of CH/CFTL and CH/CD-CFTL

Compared with CFTL and CD-CFTL in Fig. 6, the T_s, T_d, and T_p of CH/CFTL and CH/CD-CFTL has very little change. For CH/CFTL, the reason is that CH adsorbed by CFTL is easily released completely in a humid and hot environment. For CH/CD-CFTL, the reason is that the inclusion-release reaction is an equilibrium reaction.

Although the inclusion compound has higher stability, CH could be completely released under certain conditions. However, the inclusion-release reaction could be used to prepare antibacterial, sweet-scented, and other functional materials.

3.3 Antibacterial properties

The characterization of CH/CD-CFTL proved that the antibacterial agent has been included in the cavities of β-CDs on CD-CFTL, so theoretically it should have better antibacterial properties. The gram-negative bacteria *E. coli* and the gram-positive bacteria *S. aureus* were selected as the test bacteria and the antibacterial properties were investigated by the natural colony counting method using the oscillation method.

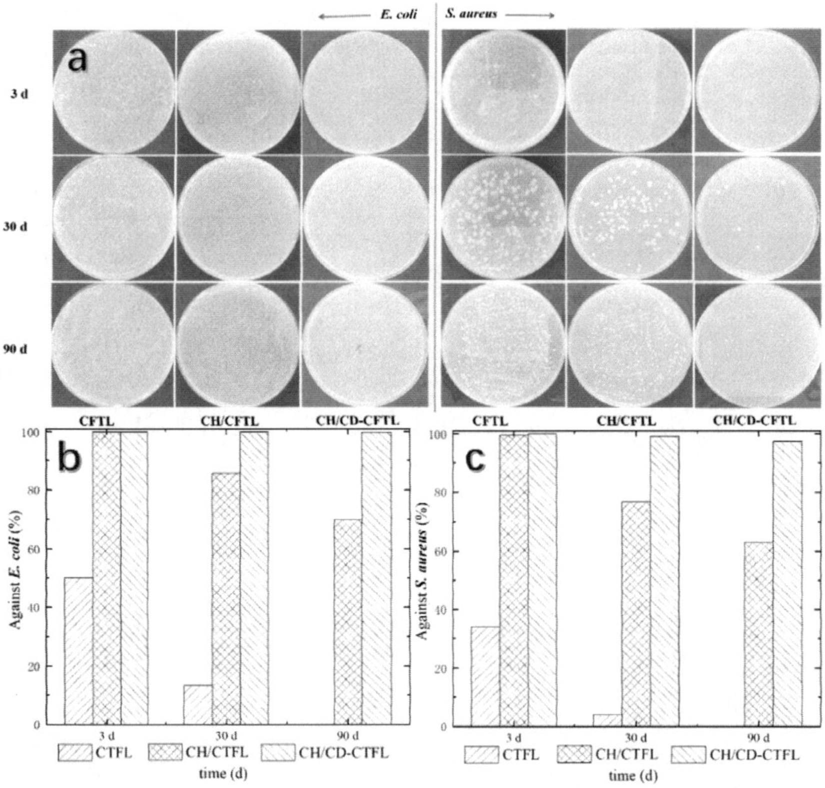

Fig. 10 Against *E. Coli* and *S. aureus* for 3 days, 30 days, and 90 days

The antibacterial results of CFTL, CH/CFTL, and CH/CD-CFTL after being stored in natural conditions for 3 days, 30 days, and 90 days were shown in Fig.10a, and the statistical results were shown in Fig. 10b and Fig. 10c. CFTL had slight antibacterial properties, which might be related to the used tannin extract and organic tanning agent. But after 30 days, CTFL had a poor antibacterial ability, which could not meet the antibacterial requirements. CH/CFTL has a certain antibacterial effect, however, the antibacterial effect decreases with the increase of time. The reason might be the weak force between CH and CFTL, and the low usage of CH (0.05%) which was only half of the conventional minimum dosage.

Different from CH/CFTL, CH/CD-CFTL has excellent long-lasting antibacterial properties, and the antibacterial rates after 90 days of storage against *E. coli* and *S. aureus* still reached 99.8% and 97.4%, respectively, which were 29.9% and 34.4% higher than those without β-CD treatment. It showed that the CH included in the cavity of β-CD had a slow-release effect and could prolong the antibacterial effect. At the same time, the antibacterial effect of *E. Coli* was better than that of *S. aureus*, which might be related to the performance of the antibacterial agent itself.

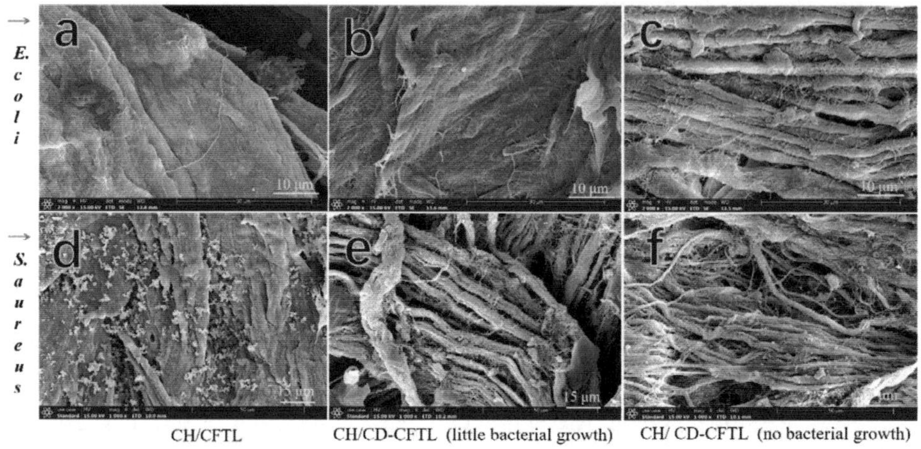

Fig.11 SEM against *E. coli* and *S. aureus* of CH/CFTL and CH/CD-CFTL after 90 days

Fig. 11 showed the SEM of the antibacterial test of CH/CFTL and CH/CD-CFTL after 90 days of storage. The *E. coli* and *S. aureus* grew wildly on samples of CH/CFTL, caused serious damage to the surface. But there were almost no bacteria on the surface of CH/CD-CFTL, and occasionally a few bacteria that have been inhibited and unable to grow were found. In summary, the obtained leather has an excellent long-term antibacterial ability.

4. Conclusion

An antibacterial strategy to improve the long-term antibacterial effect of CFTL had been proven to be successful, that is, to introduce a complete *β*-CD cavity on the CFTL and to fill the cavity with a small molecule antibacterial agent such as CH. The optimized CD-CFTL preparation process is 5% CDAD, the pH and time of the permeation reaction are 3.5 and 4 hours, the binding reaction is 3.5 and 4 hours, and the reduction reaction time is 2 hours, which can make the CD grafting rate reach

2.51%. The optimized CH/CD-CFTL preparation process is 0.05% CH addition, the pH, temperature, and time of inclusion reaction are 4.0, 30℃ and 4 hours, which can make the inclusion rate of CH and the CH loading degree of CD-CFTL reach 83.04% and 0.42 mg/g, respectively. It should be noted that CFTL is based on the combined tanning of organic tanning agent F-90, tannin extract and Zr-Al-Ti complex tanning agent, and CDAD is obtained by oxidizing the primary hydroxyl group at the C_6 position of β-CD, which has the complete β-CD cavity structure.

The structure and properties of CD-CFTL and CH/ CD-CFTL were characterized by FTIR, SEM, DSC, TGA, and the antibacterial rate against *E. coli* and *S. aureus* was calculated according to the oscillation method. The results showed that the T_s, T_d, and T_p increased 5.7 °C, 4.9 °C, and 6.9 °C, respectively. After being treated by CH with 90 days of storage, the antibacterial rates against *E. coli* and *S. aureus* still reached 99.8% and 97.4%, respectively, which were 29.9% and 34.4% higher than those without β-CD treatment.

Since both antibacterial properties and heat resistance could meet the requirements of production applications, the application prospects were promising. In conclusion, a new antibacterial method could improve the long-term antibacterial performance and heat resistance of the chrome-free tanned leather, and be expected to be applied to the leather industry.

Declarations

Competing interests

The authors declare that they have no competing interests.

Funding

This work was financially supported by the Fujian Provincial Natural Science Foundation Projects [2019J06025], the National Key R&D Program [2017YFB0308500], and the Fundamental Research Funds for the Central Universities [20826041E4156].

Acknowledgements

The authors gratefully acknowledge the Xingye Leather Technology Co., Ltd for their providedthe wet salt hides, and would thanks to Prof. Hui Wang (Analytical&Testing Center, Sichuan University) for her help in SEM observations.

References

Barros M. C., Fernandes I. P., Pinto V., et al., (2011): Chitosan as an antimicrobial agent for footwear leather components. Biodegradable Polymers and Sustainable Polymers (BIOPOL-2009), 151-166.

Bowes J. H., Cater C. W., (1968): The interaction of aldehydes with collagen., Biochimica et Biophysica Acta (BBA) - Protein Structure., 168(2), 341-352.

Chao D., Qian L., Zhao G., et al., (2014): Preparation of antimicrobial cellulose fibers by grafting β-cyclodextrin and inclusion with antibiotics., Materials Letters., 124, 181-183.

Chao J., Chen L., Hao X., et al., (2002): Preparation and study on the solid inclusion complex of ciprofloxacin with β-cyclodextrin., Spectrochimica Acta Part A: Molecular and Biomolecular Spectroscopy., 58(13), 2809-2815.

Choi J. M., Park K., Lee B., et al., (2017): Solubility and bioavailability enhancement of ciprofloxacin by induced oval-shaped mono-6-deoxy-6-aminoethylamino-beta-cyclodextrin., Carbohydrate Polymer, 163, 118-128.

Cuadros S., Manresa M. A., Font J., et al., (2012): Alternative Fungicides for the Leather Industry: Application in Various Processes., 96(6), 225-233.

Eastburn S. D., Tao B. Y., (1994): Applications of modified cyclodextrins. Biotechnology Advances, 12(2), 325-339.

Gauzit A. A., Palomino D. C., Maton M., et al., (2020): Designed sponges based on chitosan and cyclodextrin polymer for a local release of ciprofloxacin in diabetic foot infections., International Journal of Pharmaceutics., 587, 119677.

Geng X., Li W., Yin Q., et al. , (2018): Design and fabrication of reversible thermochromic microencapsulated phase change materials for thermal energy storage and its antibacterial activity., Energy, 159, 857-869.

Hou D., Wang R., Wang Z., et al., (2021): A light-activatable antibiotic with high activation efficiency and uncompromised bactericidal potency in the activated state. Journal of Leather Science and Engineering, 3, 8.

Hu J., Ye C., Zhao Y., et al., (1999): A Facile Synthesis of β-Cyclodextrin Monoaldehyde., Chinese Chemical Letters., 10, 273-274.

Kanagaraj J., Panda R. C., Vinodh K. M., (2020): Trends and advancements in sustainable leather processing: Future directions and challenges - A review. Journal of Environmental Chemical Engineering., 104379.

Krishnamoorthy G., Sadulla S., Sehgal P. K., et al., (2012): Green chemistry approaches to leather tanning process for making chrome-free leather by unnatural amino acids. Journal of Hazardous Materials., 215, 173-182.

Liu H., (2008): Study on changes of the IR absorption bands for the Guest Molecule in β-cyclodextrin inclusion complex., Spectroscopy and Spectral AmIySis., 28(10), 91-92.

Matencio A., Navarro-Orcajada S., García-Carmona F., et al., (2020): Applications of cyclodextrins in food science. A review., Trends in Food Science & Technology, 104, 132-143.

Orlita A., (2004): Microbial biodeterioration of leather and its control: a review. International Biodeterioration & Biodegradation., 53(3), 157-163.

Radu C. D., Parteni O., Ochiuz L., (2016): Applications of cyclodextrins in medical textiles — review. Journal of Controlled Release, 224, 146-157.

Sathish M., Madhan B., Sreeram K J., et al., (2016): Alternative carrier medium for sustainable leather manufacturing – a review and perspective. Journal of Cleaner Production., 112, 49-58.

Valle D, Martin E M., (2004): Cyclodextrins and their uses: a review., Process Biochemistry, 39(9), 1033-1046.

Velmurugan P, Cho M, Lee S M, et al., (2014): Antimicrobial fabrication of cotton fabric and leather using green-synthesized nanosilver. Carbohydr Polymer., 106, 319-325.

Wen H., Chen J., Yang Y., et al., (2021): Redesign and application of chrome-free tanning and its performance endowed process technology., China Leather., 50(11), 1-7.

Wen H., Yu G., Zheng X., et al., (2021): Improving the antibacterial properties of wet leather based on grafting β-cyclodextrin and inclusion with ciprofloxacin hydrochloride., Journal of Society of Leather Technologists and Chemists., 104(5): 1-14.

Wen H., Zheng X., Yu G., et al., (2021): Preparation of fragrant collagen fibers by cross-linking with β-cyclodextrin and inclusion of l-menthol: characterization, release behavior and reusability evaluation. The Journal of The Textile Institute., doi: 10.1080/00405000.2021.1992856: 1-10.

Xiao S., Zhu J., Dan W., et al., (2014). Application of n-(2-hydroxy) propyl-3-trimethyl ammonium chitosan chloride as an antibacterial and antifungal agent in wet-blue cattle hides. Journal of the Society of Leather Technologists and Chemists., 98(6), 264-268.

Yap K. L., Liu X., Thenmozhiyal J. C., et al., (2005): Characterization of the 13-cis-retinoic acid/cyclodextrin inclusion complexes by phase solubility, photostability, physicochemical and computational analysis [J]. Eur J Pharm Sci, 2005, 25(1): 49-56.

Yu Y., Wang Q., Yuan J., et al., (2016): A novel approach for grafting of beta-cyclodextrin onto wool via laccase/TEMPO oxidation., Carbohydrate Polymer., 153, 463-470.

Zhang H., Hou X., Liang L., (2012): Study on the supramolecular systems of ciprofloxacin with cyclodextrins., Journal of Molecular Science., 28(3): 214-220.

Zhao H., Heindel N. D. , (1991): Determination of Degree of Substitution of Formyl Groups in Polyaldehyde Dextran by the Hydroxylamine Hydrochloride Method., Pharmaceutical Research., 8(3), 400-402.

Zhu X., (2008): Thermal Stability and Thermal Decomposition Kinetics of Ofloxacin., Chemical World., 19(6), 333-334.

VEGETABLE TANNINS USES AND ANTIOXIDANT PROPERTY

Lívia de Souza Schaumlöffel[a,*], Luiz Antonio Mazzini Fontoura[b], Mariliz Gutterres[a]

[a]Federal University of Rio Grande do Sul (UFRGS), Chemical Engineering Post-Graduated Program (PPGEQ), Laboratory for Leather and Environmental Studies (LACOURO)

[b]Lutheran University of Brazil

[*]Corresponding author. Address: Eng. Luiz Englert str., s/n, Postal Code 90040-040, Porto Alegre, RS, Brazil Tel.: +55 51 3308 3638. E-mail: liviass@ufrgs.br

Abstract

Vegetable tannins are phenolic compounds of great economic and ecological interest. They are found in roots, wood, bark, leaves, fruits, seeds and sap of different plant species. The tannins composition and their chemical characteristics that allow a series of reactions to modify their chemical structure give possibilities for a wide range of uses for them. Historic used as tanning agents, recently, the interest for different applications of tannins has grown and several studies evaluated either they can be successfully used or not as flocculants, dispersants, antioxidants, biocides, etc. The antioxidant properties of tannins are well known and they could be used with this purpose in foods, beverages, fuels, lipids systems, polymers, etc, replacing synthetic and other less available plant extracts. However, their antioxidant properties are scarcely studied. The antioxidant property of acacia (*Acacia mearnsii*), chestnut (*Castanea sativa*) and quebracho (*Schinopsis sp.*) of commercial tannins extracts were tested. The commercial tannins were also characterized by their ash content, by TG/DTA analyses, by Folin-Ciocalteau method for the determination of Total Phenolic Content, and against DPPH (1,1-diphenyl-2-picryl-hydracyl) radical for antioxidant activity. The results showed that plant tannins reach a maximum of 87% DPPH radical inhibition at a concentration of 10 mg L^{-1}.

Keywords: tannins, antioxidant, chromium (VI)

1 Introduction

Vegetable tannins are phenolic compounds found in the roots, wood, bark, leaves, fruits, and seeds of different plant species where they are synthesized as secondary metabolites with protective function against insects, fungi, or bacteria. Chemically, tannins are complex polyphenolic substances and, according to their structures, they are divided into two major groups: hydrolysable tannins and condensed tannins. The main part of the composition comes from the structure of the monomeric units: catechin and catechin-based compounds (flavonols, galoflavanols, and flavanol gallate) for condensed tannins; gallic acid and derivatives (ellagic, hexahydroxydiphenic acids) for hydrolysable tannins (Das et al., 2020).

The content and species of tannins vary from one plant to another and also from one part of the plant to another. Hydrolyzable tannins have low availability and limited production worldwide (they account for less than 10% of the world production of commercial tannins), therefore they have a high price, which makes them less attractive. On the other hand, condensed tannins are equivalent to more than 90% of the world production (García et al., 2016).

1.1 Vegetable tannins applications

Tannins extracts are compounds of great economic and ecological interest since the diversity of structures found in the different tannins allows a variety of uses for them. Brief research in recent literature demonstrates the growing interest in vegetable tannins uses. A search in the Scopus database by the words "vegetable and tannins" found 1885 documents in a forty years period (Figure 1). The studies cover several applications in different fields, many studies cover improvement of already commercial uses and others are in the research stage.

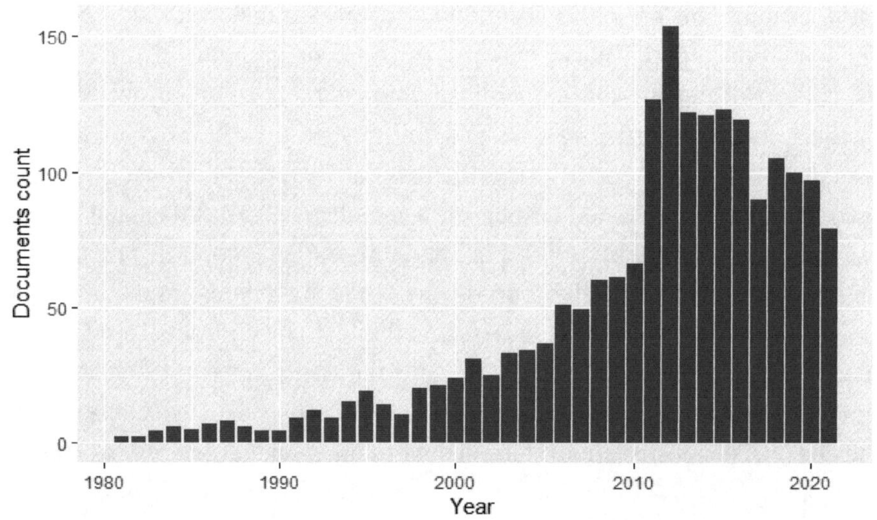

Figure 1. Publications with the terms 'vegetable' and 'tannins' in the Scopus Database

Among the industrial uses of tannins, the most traditional and common application is leather tanning. It is common for this application that the tannins undergo a sulfitation process, to facilitate the extraction process, whitening of the extract and improvement in the tanning properties (Grasel et al., 2016).

The second most important application is being adhesives for wood products, where the tannins are the more interesting alternative to synthetic adhesives. The growing interest in this application is due that tannin-based adhesives are less prone to emitting formaldehyde, a carcinogenic compound. Further, studies are also focusing on the tannins ability to react with hexamine, used as a hardener, which can produce an adhesive not prone to decompose to formaldehyde at all (Shirmohammadli et al., 2018).

The use of tannins in water treatments is also widespread. In this application, vegetable tannins represent natural alternatives to inorganic metal-based flocculants/coagulants and the use of tannins as flocculant agents is known since 1970, however, for a long time, this use was neglected. The interest for tannins as flocculants/coagulants began to grow again near 2009 due to the concerns for sustainability and the replacement of chemicals by bio-based alternatives (Pizzi, 2019).

Among other industrial applications for tannins extracts are their use as wood preservatives, corrosion inhibitors and foams. Table 1 present a resume of the important property related to each application.

An important characteristic of tannins is related to the OH groups that confer antioxidant and anti-radical properties to these compounds. Phenolic groups donate hydrogen, which results in the formation of stable free radicals due to the occurrence of resonance in the aromatic ring and, therefore, hinders free radical oxidation chain reactions (Varatharajan & Pushparani, 2018). These antioxidants properties are relevant in many fields.

In the beverage industry, tannins are added in the winemaking process to improve several properties, such as colour stability, odour and mouth-feel, flavour and also promote protections against oxidation and proteic material precipitation. The tannins are naturally present in grapes and were added in the ageing process in oak barrels. However, they can be added directly reducing the ageing time and also compensating low tannin levels in some wines. Tannins used in winemaking are known as "enological tannins" (Laghi et al., 2010).

Their application as food additives is also of great interest. Tannins are used in animal food due to their antioxidants abilities and researches findings relating their use with the improvement of animal digestion. The use in foods and beverages also beneficiate from the know bactericide ability of tannins (Fraga-Corral et al., 2020).

The antibacterial property, together with the antioxidant abilities raise the interest for medicinal and farmaceutical use of tannins where many applications, such as antiviral, anti-fungal and anti-cancer, are under research (Fraga-Corral et al., 2020; Pizzi, 2019).

Especially interesting is the ability of tannins to prevent Cr (VI) formation, which may be also related to the antioxidants properties (Fuck et al., 2011; Ozkan et al., 2015).

Table 1. Properties related to each application of vegetable tannins extracts

Application	Relevant property
Tanning agent	Ability to complex proteins: tannin polyphenols can create crosslinking with collagen by numerous hydrogen bonds. At first, hydroxyl groups (OH) bind with the active collagen centres, and then continue to fill the interfibrillar spaces in leather.
Adhesive	Cross-linking of hydroxyl groups of tannins with formaldehyde and others compounds.
Flocculant	Adjacent phenolic hydroxyls display a specific affinity to metal ions providing the ability to chelate them.
Corrosion inhibitor	The phenol groups oxides and then complex with the metal by ortho-diphenol complexation, protecting from external oxidation.
Antioxidant	OH groups stabilize free radicals and then inhibit further damage these may otherwise cause.
Bactericide	Tannins react with proteins irreversibly, thus complexing within bacterial membranes, neutralizing their activity.

Research studies in the Laboratory for Leather and Environmental Studies (LACOURO) covered a few of these uses and brought new knowledge for each specific application. A resume of the findings, among others in the literature, is shown in Table 2.

Table 2. Recent studies in vegetable tannins applications

Application	Objective	Conclusions	Reference
Tanning agent	Study relevant chemical parameters and the correlation with their cross-sectional penetration in the hide	Myrobalan and tara tannins presented low penetration potential while acacia, quebracho, and chestnut went through the skin. The properties associated with the penetration in the hide were the higher tannin content and lower insoluble content present in the material.	(Auad et al., 2019)
Wastewater treatment	Reduce aluminium salts in the coagulation-flocculation process	The values of colour removal, turbidity and total organic carbon were statistically equal to the application with and without tannin. Therefore, is possible to use aluminium sulphate with tannins in order to reduce the	(Anjos, 2016)

		dosages of aluminium sulphate.	
	Recovery of microalgae biomass growth in the wastewater	There was a positive influence in the use of vegetable tannins for the sedimentation of microalgae throughout the neutralization of the impingent charges.	(Fontoura, 2017)
Adhesive	Development of a cornstarch-tannin-sugar adhesive with zero formaldehyde for interior particleboard	Tannin-based resin satisfied the exigencies of panels for interior fittings used in dry medium (P2) according to European norms EN 312 (2010)	(Oktay et al., 2021)
Corrosion inhibitor	Tannins as corrosion inhibitor for AA1200 aluminium alloys	Results show that natural tannin is an efficient inhibitor coated for aluminium Alloy 1200	(Nardeli et al., 2019)
Antioxidant (prevent biofuel degradation)	Evaluate tannin-based antioxidant in the preservation of biodiesel	The protection of oxidation by tannin antioxidant was superior to the synthetic TBHQ (terc-butil-hidroquinona)	(Schaumlöffel et al., 2021)
Antioxidant (preventing Cr (VI) formation)	Evaluate the effects of tannins in Cr(VI) formation in leather	The use of 2.5 and 5.0% of any type of vegetable tannin prevented Cr(VI) formation in all leather samples	(Ozkan et al., 2015)
	Evaluate the Cr(VI) formation in leather in different retanning conditions in leather	The retanning with vegetable tannin has a beneficial effect by preventing the oxidation of chromium	(Fuck et al., 2011)

An important aspect to be considered in the broad applications of vegetable tannins is their biodegradability. Their anti-microbiologic qualities may cause problems at end of live cycles where the tannins-based products may undergo microbiological degradation. The work of Agustini et al. (2018) demonstrates that chromium sludge was a more suitable medium for anaerobic digestion than vegetable tannin sludge. They identified it in all parameters studied (BOD_5, TOC, VSS, NT, IC, pH). Another study evaluated the influence of the concentration of tannin, the forms of culture and the type of tannin used in the cultivation of *Aspergillus niger* in solid-state and submerged culture cultivations (Spier & Gutterres, 2018). The consumption of tannins in solid-state and submerged cultivation was similar at 13% and 15%, respectively proving the fungistatic action of the phenolic compounds present in the tannins on *Aspergillus niger*. However, the same study presents an alternative treatment in which fungal strains of the genus Penicillium were more effective reaching 97.3% consumption of acacia tannin.

With this overview, it can be seen that many different factors are relatable to tannins application, thus, the objective of this study was to evaluate some tannins properties, relating them to new applications under research, especially tannins use as antioxidants.

2 Materials and methods

2.1 Materials

The tannins evaluated were the commercial tannins extracts of acacia (Tanac S.A.), quebracho (Silvateam S.A.), and chestnut (Silvateam S.A.).

2.2 Tannins characterization

The determination of total phenolic content (TPC) was carried out by the Folin-Ciocalteau method. For the TPC determination, a 0.1 mL aliquot of tannin extracts stock solution (160 ppm) was mixed with 0.4 mL of ethanol solution (10% v/v in distilled water) and 2.5 mL of the Folin-Ciocalteau reagent (10% v/v in distilled water). After 3 min, 2.0 mL of 7.5% (m/v) Na_2CO_3 solution was added and the mixture remained at rest from 1h, protected from light. The absorbance was measured at 760 nm in the UV–Vis spectrophotometer (Thermo Scientific Visible Spectrophotometer – GENESYS 30). The TPC was determined using a standard curve of gallic acid and the results were expressed in gallic acid equivalent per gram of sample (mg EAG/g).

The thermal characterization of the compounds was performed in a thermogravimetric analyzer (TA Instruments - SDT Q600) within the 25–900 °C range and a 10 °C/min heating rate, under a nitrogen atmosphere.

Ash content was determined by TAPPI T211 standard method.

2.3 Antioxidant activity analysis

Antioxidant activity was determined by the 2,2-diphenyl-1-picryl-hydrazine (DPPH) radical scavenger method. Briefly, 3.7 mL of DPPH radical solution in methanol (40 mg L−1) were mixed with 0.3 mL of tannin extracts and TBHQ samples, with concentrations varying from 67 to 200 mg/L. The samples remained at rest in the dark for 30 min. Subsequently, the absorbance was measured at 517 nm (Thermo Scientific Visible Spectrophotometer – GENESYS 30). The percentage of DPPH radical activity inhibition (percentage of antioxidant activity) was calculated by Eq. (1),

$$(\%)\ inhibition = \frac{(A0 - At)}{A0} \times 100 \qquad (1)$$

where A0 is the absorbance of the pure DPPH solution and At the absorbance of DPPH in the presence of a sample.

3 Results

3.1 Tannins characterization

The total phenol content found (presented in Table 3) evidenced a large amount of phenols present in the vegetable tannins, especially when compared to other plant extracts (Schaumlöffel et al., 2021). Thus, applications that depend on this structure, such as antioxidant and corrosion inhibitors, can benefit from the use of vegetable tannins.

Table 3. Ash and total phenolic content for the different tannins.

Tannin	**TPC (mgEAG g^{-1})**	**Ash content (%)**
Acacia	645.2 $\pm$ 21.5	3,13 ± 0,16
Chestnut	511.6 $\pm$ 115.2	1,40 ± 0,18
Quebracho	660.2 $\pm$ 44.8	8,35 ± 0,25

The ash content is related to the salt plants absorb while growing and to the salts added in the sulfitation process (Weiler et al., 2019). Although due to the origin of ashes their content may vary from a tannin sample to another, the ash values found (Table 3) are similar to those in other studies (Zanetti et al., 2021). This amount can be considered small in the case of applications that require full combustion at the end of its use, as is the case of the combustion of wastewater sludge to energy production. However, in applications where the final product specification requires low ashes, such as the use as an antioxidant in oils, food and beverages, the ash content should be considered when the tannins are added in great quantity, especially the quebracho tannin.

The thermal analyses (Figure 2) demonstrate that the tannins lost about 10% of weight in 143, 134, 157 °C for acacia, chestnut and quebracho, respectively. Near 80% of weight still present for of acacia and quebracho at 250 °C and for chestnut at 295 °C. These are indications that applications that require exposition to higher temperatures require evaluation of thermal behaviour.

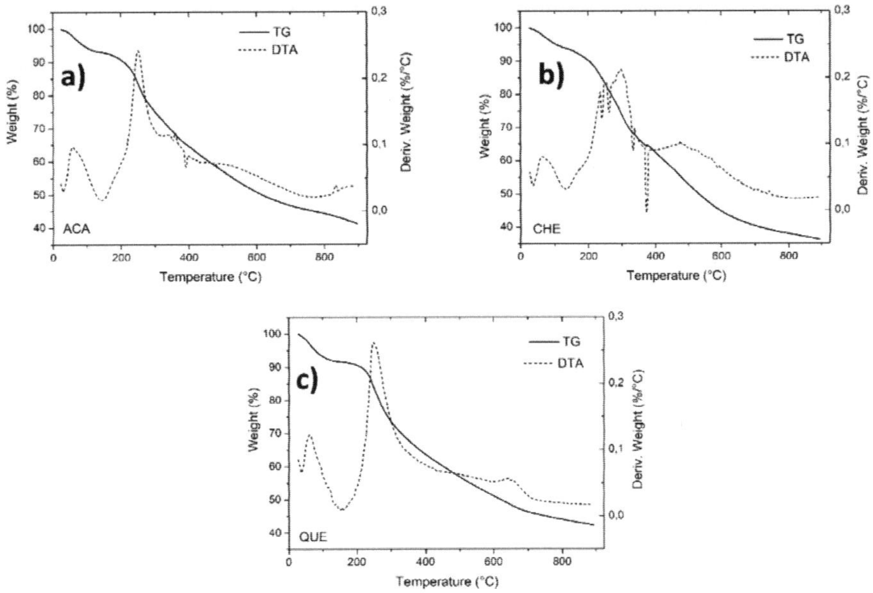

Figure 1. TG/DTA curves for ACA, CHE and QUE tannins

3.2 *Antioxidant ability*

The antioxidant capacity was evaluated by the DPPH radical scavenger essay and the results are shown in Figure 3. The tannins with a concentration of 10 mg/L were able to inhibit more than 85% of DPPH radical in aqueous medium. Compared with other vegetable extracts like green tea extracts, that inhibited 70% with a concentration of 250 ppm (Bharti & Singh, 2020), tannins effects present superior performance.

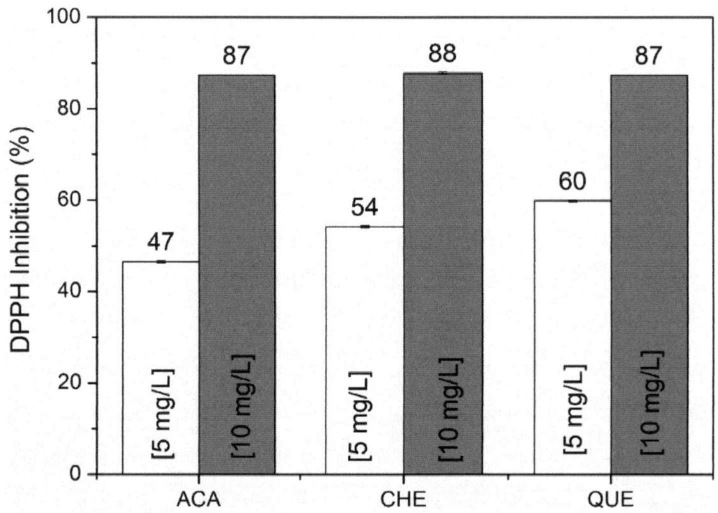

Figure 3. DPPH Scavenging activity of acacia (ACA), chestnut (CHE) and quebracho (QUE) tannins

4 Conclusion

Vegetable tannins are natural compounds, therefore they are been considered as an alternative material for replacing synthetic compounds in many applications, providing more sustainable processes. Their properties and structure make them suitable for a broad range of applications, such as tanning agent, adhesive, flocculant, corrosion inhibitor, antioxidant, food additive and for medicinal uses. The analyses of some tannins properties showed high total phenolic content. The ashes content, on the other hand, were small. The thermal analyses showed that at 250°C 80% of tannins mass were still present and the antioxidant activity results showed that vegetable tannins reach 87% of DPPH inhibition at a concentration of 10 mg/L.

These analyses demonstrate the potential for the use of tannins as antioxidants due to their phenolic content and antioxidant activity. However, further studies are required to scale new research findings in industrial applications, which can increase the competitiveness of tannins productive chains.

5. Acknowledgments

We would like to thank the scholarships provided by Brazilian Federal Agency for the Scholarship (CAPES).

References

Agustini, C. B., Meyer, M., Da Costa, M., & Gutterres, M. (2018). Biogas from anaerobic co-digestion of chrome and vegetable tannery solid waste mixture: Influence of the tanning agent and thermal pretreatment. *Process Safety and Environmental Protection*, *118*, 24–31.

Anjos, P. S. dos. (2016). *Aplicação de sulfato de alumínio e taninos vegetais na coagulação-floculação de efluentes de curtume.*

Auad, P., Spier, F., & Gutterres, M. (2019). Vegetable tannin composition and its association with the leather tanning effect. *Chemical Engineering Communications*, 1–11.

Bharti, R., & Singh, B. (2020). Green tea (Camellia assamica) extract as an antioxidant additive to enhance the oxidation stability of biodiesel synthesized from waste cooking oil. *Fuel*, *262*.

Das, A. K., Islam, M. N., Faruk, M. O., Ashaduzzaman, M., & Dungani, R. (2020). Review on tannins: Extraction processes, applications and possibilities. *South African Journal of Botany*, *135*, 58–70.

Fontoura, J. T. da. (2017). *Crescimento de microalgas em efluente de curtume : remoção de nutrientes , viabilidade de produção de biocombustíveis e utilização da biomassa residual* [Tese (Doutorado em Engenharia Química)]. Escola de Engenharia, Universidade Federal do Rio Grande do Sul.

Fraga-Corral, M., García-Oliveira, P., Pereira, A. G., Lourenço-Lopes, C., Jimenez-Lopez, C., Prieto, M. A., & Simal-Gandara, J. (2020). Technological application of tannin-based extracts. *Molecules*, *25*(3).

Fuck, W. F., Gutterres, M., Marcílio, N. R., & Bordingnon, S. (2011). The influence of chromium supplied by tanning and wet finishing processes on the formation of Cr(VI) in leather. *Brazilian Journal of Chemical Engineering*, *28*(2), 221–228.

García, D. E., Glasser, W. G., Pizzi, A., Paczkowski, S. P., & Laborie, M. P. (2016). Modification of condensed tannins: From polyphenol chemistry to materials engineering. *New Journal of Chemistry*, *40*(1), 36–49.

Grasel, F. D. S., Ferrão, M. F., & Wolf, C. R. (2016). Development of methodology for identification the nature of the polyphenolic extracts by FTIR associated with multivariate analysis. *Spectrochimica Acta - Part A: Molecular and Biomolecular Spectroscopy*, *153*, 94–101.

Laghi, L., Parpinello, G. P., Rio, D. Del, Calani, L., Mattioli, A. U., & Versari, A. (2010). Fingerprint of enological tannins by multiple techniques approach. *Food Chemistry*, *121*(3), 783–788.

Nardeli, J. V., Fugivara, C. S., Taryba, M., Pinto, E. R. P., Montemor, M. F., & Benedetti, A. V. (2019). Tannin: A natural corrosion inhibitor for aluminum alloys. *Progress in Organic Coatings*, *135*, 368–381.

Oktay, S., Kızılcan, N., & Bengü, B. (2021). Development of bio-based cornstarch - Mimosa tannin - sugar adhesive for interior particleboard production. *Industrial Crops and Products*,

170, 113689.

Ozkan, C. K., Ozgunay, H., & Kalender, D. (2015). Determination of antioxidant properties of commonly used vegetable tannins and their effects on prevention of Cr(VI) formation. *Journal of the Society of Leather Technologists and Chemists*, *99*(5), 245–249.

Pizzi, A. (2019). Tannins: Prospectives and actual industrial applications. *Biomolecules*, *9*(8), 344.

Schaumlöffel, L. S., Fontoura, L. A. M., Santos, S. J., Pontes, L. F., & Gutterres, M. (2021). Vegetable tannins-based additive as antioxidant for biodiesel. *Fuel*, *292*, 120198.

Shirmohammadli, Y., Efhamisisi, D., & Pizzi, A. (2018). Tannins as a sustainable raw material for green chemistry: A review. *Industrial Crops and Products*, *126*, 316–332.

Spier, F., & Gutterres, M. (2018). Biodegradation by filamentous fungi of vegetable tannins used in the tanning industry. *Journal of the Society of Leather Technologists and Chemists*, *102*(2), 59–63.

Varatharajan, K., & Pushparani, D. S. (2018). Screening of antioxidant additives for biodiesel fuels. *Renewable and Sustainable Energy Reviews*, *82*, 2017–2028.

Weiler, F. H., dos Santos Grasel, F., Vescia Lourega, R., da Silva Ramos, A., & Flôres Ferrão, M. (2019). Simultaneous determination of sulfur, nitrogen and ash for vegetable tannins using ATR-FTIR spectroscopy and multivariate regression. *Microchemical Journal*, *149*, 103994.

Zanetti, M., Cesprini, E., Marangon, M., Szczurek, A., & Tondi, G. (2021). Thermal valorization and elemental composition of industrial tannin extracts. *Fuel*, *289*, 119907.

AUTHOR INDEX